# Global Decline of Insects

*Edited by Hamadttu Abdel Farag El-Shafie*

Published in London, United Kingdom

IntechOpen

*Supporting open minds since 2005*

Global Decline of Insects
http://dx.doi.org/10.5772/intechopen.94711
Edited by Hamadttu Abdel Farag El-Shafie

Contributors
Barkat Hussain, Showket A. Dar, Mohmmad Javed Ansari, Yahya Al Naggar, Syed Nighat, Syed Burjes Zehra, Rizwan Rashid, Mudasir Hassan, Farkhanda Manzoor, Mahnoor Pervez, Gagan Preet Kour Bali, Amritpal Singh Kaleka, Toheed Iqbal, Kiran Shahjeer, Nazeer Ahmed, Saeed Ahmed, Khalid Awadh Al-Mutairi Al-Mutairi, Reham Fathey Ali, Hanem Fathy Khater, Muzafar Riyaz, Rauf Ahmad Shah, Soosaimanickam Maria Packiam, Thenepalli Sudha Rani, Potireddy Suvarna Latha Devi, Mukhtar Alam, Muhammad Saeed, Hidayat Ullah, Rafi Ullah, Nibal Abd Aleem Hassan Ahmed, Hanem Fathy Khater, Muhammad Salman, Shireen Saleem, Shoeba Binte Anis, Mumuni Abudulai, Jerry Asalma Nboyine, Ahmed Seidu, Peter Quandahor, Fousséni Traore, Navkiran Kaur, Hamadttu Abdel Farag El-Shafie



First published in London, United Kingdom, 2022 by IntechOpen
IntechOpen is the global imprint of INTECHOPEN LIMITED, registered in England and Wales, registration number: 11086078, 5 Princes Gate Court, London, SW7 2QJ, United Kingdom
Printed in Croatia

British Library Cataloguing-in-Publication Data
A catalogue record for this book is available from the British Library

Additional hard and PDF copies can be obtained from orders@intechopen.com

Global Decline of Insects
Edited by Hamadttu Abdel Farag El-Shafie
p. cm.
Print ISBN 978-1-83969-587-2
Online ISBN 978-1-83969-588-9
eBook (PDF) ISBN 978-1-83969-589-6

For EU product safety concerns:
IN TECH d.o.o., Prolaz Marije Krucifikse Kozulić 3, 51000 Rijeka, Croatia, info@intechopen.com or visit our website at intechopen.com.

# Meet the editor

Hamadttu Abdel Farag El-Shafie is an Associate Professor of Entomology and a senior research entomologist at the Date Palm Research Center of Excellence, King Faisal University, Saudi Arabia. He is the head of the sustainable pest management research program at Date Palm. He obtained his BSc and MSc from the University of Khartoum, Sudan, in 1988 and 1993, respectively. He received his Ph.D. from the University of Giessen, Institute of Phytopathology and Applied Entomology, Germany, in 2001. In 2008, Dr. El-Shafie was appointed head of the Crop Protection Department, and then deputy dean for academic affairs at the Faculty of Agriculture, University of Khartoum, Sudan. He supervised twenty-five MSc students and five Ph.D. students at the University of Khartoum. His research interest focuses on the sustainable management of field crop pests using biopesticides and semiochemical-based technologies. He has more than twelve years of experience in the management of the invasive red palm weevil and other date palm pests of major significance. He has published seventy research papers in international peer-reviewed journals and seventeen book chapters with international publishers. He has also edited two books. Dr. El-Shafie has participated in more than thirty international conferences in the field of entomology. During the last decade, he has been reviewing manuscripts for thirty-five renowned international journals.

# Contents

# Preface

Insects represent a diverse group of animals with numerous species on the planet. They contribute significantly to maintaining and proper functioning of different ecosystems. Insects pollinate crops, improve soil characteristics, provide food for other animals, recycle nutrients, and control insect pests. During the last decade, scientists began to speak loudly about global insect decline. They used alarming terminologies such as defaunation, insectageddon, insect apocalypse, and extinction in the literature to describe such decline in insect abundance, biomass, and species richness. The use of such alarming words may be justified in order to raise awareness about the negative impacts of insect decline on agricultural sustainability, the environment, and food security. The decline of insects could severely affect birds, reptiles, amphibians, fish, and small mammals that utilize insects as a source of food. This will have great repercussions on the overall ecosystem. Evidence shows that the numbers of different groups of insects including butterflies, moths, bumblebees, stingless bees, honeybees, dragonflies, and beetles are beginning to decrease. There are many reasons for this global decline, including agricultural intensification, urbanization, habitat destruction, and climate change. However, intensive agriculture, particularly the heavy and unwise use of pesticides that persist in the ecosystem, is the major cause.

This book amalgamates information pertaining to the global decline of insects with emphasis on the potential reasons behind this decline and the possible means of mitigating it. It contains eleven chapters distributed into two sections. Section 1 includes five chapters that discuss potential causes of insect decline. Section 2 contains six chapters that elaborate on potential measures to mitigate this decline.

Chapter 1 by Dar et al. discuss the causes and reasons for insect decline with emphasis on factors such as heavy use of pesticides, habitat destruction, urbanization, climate change, and introduction of new invasive species. In Chapter 2, Bali and Kaleka elaborate on the systematic drivers of insect decline including habitat or landscape fragmentation, deforestation, and climate change. In Chapter 3, Abudulai et al. explain the impact of agricultural intensification on insect abundance and biodiversity. Manzoor and Pervez highlight the potential impact of pesticides on the honeybee in Chapter 4. In Chapter 5, Kaleka and Kaur provide useful information on the diversity, importance, and decline of pollinating insects in the present era. They point out the significance of the different groups of pollinators such as bees, wasps, butterflies, moths, ants, and beetles as well as the reasons behind their decline in the different ecosystems.

Section 2 contains chapters dealing with the measures that could halt or mitigate the impacts of global insect decline on the planet's ecosystems. In chapter 6, Ahmed et al. discuss plant-based biological insecticides as alternative and environmentally friendly insect control methods with a benign or mild adverse effects on insects. Likewise, in Chapter 7, Iqbal et al. address botanicals with active ingredients having insecticidal, antifeeding, and repellent properties. In Chapter 8, Thenepalli explains the role of microorganisms in the biodegradation of pesticides and the impact of the whole process in the alleviation of the harmful effect of pesticides on insects.

In Chapter 9, Muzafar et al. highlight the importance of proper insect conservation and management reducing the global decline of insect abundance and species richness. Minimal intervention in the ecosystem will help maintain insect numbers and biodiversity. In this respect, in Chapter 10, Saleem and Anis describe a new species of a natural enemy of pests, which could play an important role in checking the numbers of harmful insects. Finally, in Chapter 11, El-Shafie discusses the impact of organic farming on insect abundance and biodiversity, with organic farming as an agricultural approach for the production of food with the aim of restoring ecosystems.

This book is for entomologists, ecologists, botanists, environmentalists, students, and amateurs who love insect collection and preservation.

I would like to thank the chapter authors who contributed to the book by writing, revising, and submitting their work. Without their contributions, it would have been difficult to produce this book. Other contributors who deserve thanks and acknowledgments are the staff at IntechOpen, especially Author Service Manager Ms. Karmen Daleta and Commissioning Editor Ms. Iva Simcic for their unlimited assistance during the preparation of the book.

I wish to express my sincere and heartfelt thanks to my wife Nawal, son Ayman, and daughters Hiba, Hala, Safa, and Lojain for their continuous moral support, understanding, and encouragement during the preparation of this book. The editor appreciates and acknowledges the logistic support provided by the Date Palm Research Center of Excellence, King Faisal University, Saudi Arabia. I wish to thank all members of the Department of Crop Protection, University of Khartoum, Sudan, for their support and encouragement.

**Dr. Hamadttu Abdel Farag El-Shafie**
Research Entomologist,
Date Palm Research Center of Excellence,
King Faisal University,
Al-Ahsa, Saudi Arabia

Department of Crop Protection,
Faculty of Agriculture,
University of Khartoum,
Shambat, Sudan

Section 1

# Potential Causes of Insect Decline

Chapter 1

# Causes and Reasons of Insect Decline and the Way Forward

*Showket A. Dar, Mohmmad Javed Ansari, Yahya Al Naggar, Shafia Hassan, Syed Nighat, Syed Burjes Zehra, Rizwan Rashid, Mudasir Hassan and Barkat Hussain*

## Abstract

There are lot of reasons and causes of insect decline. The main causes of insect decline is attributed to habitat destruction, land use changes, deforestation, intensive agriculture, urbanization, pollution, climate change, introduction of invasive insect species, application of pesticides, mass trapping of insects using pheromones and light traps, pathological problems on various insects, and introduction of exotic honey bees in new areas that compete with the native bees for resource portioning and other management techniques for pest management, and even not leaving any pest residue for predators and parasitoids for their survival. The use of chemical insecticides against target or non-target organisms is major cause for insect decline. The diseases and decline of the important pollinators is still a mistry for colony collapse disorder. To overcome the cause of insect decline, various conservation techniques to be adopted and augmentation of artificial nesting and feeding structures, use of green pesticides, maintaining the proper pest defender ratio (P:D), policies and reaching to political audience at global level and other factors already discussed in the chapter may be helpful for mitigating the insect decline and especially for the pollinators, a key insect for life.

**Keywords:** insect decline, pollinators, causes, effect, mitigation measures

## 1. Introduction

Globally, scientific studies have reported a large and significant decline in insect populations since decades [1]. The policies and public concerns for insect decline is scarce and highly ignored since the dawn of civilization. Garden columnist George [2] warned the world regarding the decline of insect numbers. Insect decline is a serious threat because their abundance, numbers, diversity and the extinction of whole species. Decline in population does not mean only reduction in the insect numbers but also means the restricted graphical distribution which leads to the first step of extinction [3]. The causes of insect decline, and pattern are not uniform and vary with biotic, abiotic, and anthropogenic factors of the region [4]. Globally scientists are concerned for biodiversity loss in terrestrial and aquatic vertebrate [5, 6] and are more concerned to invertebrates. Although the number of insect fauna is huge, they have always have always been ignored by ecologists and conservationists.

 

There are about 5.5 million species of insects all over the World, out of which 90% species are still not been identified and their function in ecology is unknown. The number of insects in Germany has declined up to 75% in just 4 year [7] and also declined in Netherlands as well [8].

A meta-analysis data concerned to terrestrial insects published in Journal *Science* in 2020 reported a global insect population decline by 9% per decade [9, 10] in contrast to fresh water insects whose population is enhancing very fast at 11% per decade [11, 12]. Terrestrial insects are more vulnerable to diverse threats [1] and some of the most affected insect groups namely bees, butterflies, moths, beetles, dragon flies and damselflies [13]. Anecdotal and accurate evidence of population trend variability in particular locations has been offered with high [1] apparent abundance of insects during the twentieth century, as confirmed by recollection of windscreen phenomenon [14]. The possible causes of insect decline is due to [1] habitat destruction, intensive agriculture, extensive use of synthetic pesticides, urbanization, industrialization, species introduction, shifting agriculture, genetic engineering and climate change [1]. According to one report by World Economic Forum 23% of Earth's habitat would disappear by 2100 century [15] and conservation of habitat ensures the long-term survival of life on the planet; and its loss is identified as the main threat to 85% of species listed in the IUCN Red List. Research says that not all insect orders have been affected at same rate; and some orders need to be researched and revaluated because figures from earlier periods are often not available or standard scientific techniques have not been used to quantify their decline [16, 17]. The notion of insect decline is widespread globally and various [1] insect conservation measures have been launched to the judicious use of synthetic pesticides and other measure for habitat protection [18]. German government initiated an Action Programme for insect protection in 2018 and the British entomologists and ecologists wrote an open letter to have a focused research to know the reasons and causes of insect decline [19].

## 1.1 Background of insect decline

From the previous two centuries, insect decline has started rapidly [10]. The old recorded decline of Rocky Mountain Locust during 1902 [20] in USA and recently the scientists from the University of Helsinki warned humanity about worldwide insect decline with unpredictable consequences [21]. Civil society and policymakers have an important role for the future and collective well-being of insects and most of the insects are responsible for pollination, predation and parasitisation and waste material degradation in different ecosytems. Mitigating the impact of climate change, establishing buffer zones with high-quality and manageable portions of fertile land for protection, and changing agricultural practices to foster species coexistence, besides leaving a pest residue are all things that need to be done [22].

Ten trillion locust swarms were observed during 1975, and rapidly declined afterword's but the causes of such decline is not explored [23]. This species was declared extinct in 2014 due to its continued decline. However, the fossil records concerning to insect decline revealed that insect stretched back hundreds of millions of years with the discovery of new species and their extinctions [24]. Further, it is also observed that mass insect extinction occasionally occurred by natural phemenon like volcanic activity and meteorite impacts [25]. The insect decline was highest during Permian–Triassic extinction event followed by second highest mass extinction during Cretaceous-Paleogene era [26]. However, the insect diversity and abundance has resumed rhythm after first extinction giving birth to new species at higher rates, however still it will take millions of years to restore the extinct ones [6]. The latest human caused [1] Holocene extinction of species is growing since

20th century, and much of the extinction reports were not from arthropods, with 95% decline in anthropogenic habitats like grasslands [27]. In the case of vertebrates, the Zoological Society of London (ZSL) published a research-based opinion in 2012 that insect species are decreasing globally, with direct and indirect effects on pollination, ecosystem balance, ecosystem services, livestock, and overall food production, and may decline in the near future [28]. It is estimated that 20% of all invertebrates are in grave danger of extinction as a result of Holocene extinction. Generally in Holocene era, the species with least mobility, small size, smallest host ranges and climate sensitive are most vulnerable to extinction [29].

After decades, it has been noted that species extinction is on the decline; however, precise data is unavailable. Several research found a big gap between 1840 and future predictions. However, due to the global concern about species extinction, the German Nature Reserve gained a lot of attention in 2017 [8–30].

Several studies have found that declining insect abundance, biomass, and species richness are all signs of species extinction [31]. The reports revealed a localized decline in species-friendly factors, implying that species showed a region-wide decline but not necessarily in other areas. Moths, bees, beetles, dragonflies, damselflies, and stoneflies were the most studied insects in terms of extinction [4, 32]. All these species are affected directly or indirectly in many ways through changes in environment. Many insect species have adapted to external changes when environmental conditions change in some way; however, the majority of species fail to live in altered environments.

By the year 2019, the reports published by Entomological Society of America highlighted that the available data concerning to extinction of insects is not satisfactory and insufficient to support imminent mass extinction [33]. Extrapolated forecasts may have been stretched, and data ranges about different species have been over-hyped, underestimated, and overestimated, according to society. The decline of some groups (butterflies, bees and beetles) have been documented by European studies, while other regions have recorded an increase in species count, however the definite and clear trend of insect decline from most part of the world is not available. Due to lack of sufficient information and historical measurements about majority of insect species [34]. It is very cumbersome to assess their long-term shift in abundance, diversity, and habitat. For many of the species, an exaggerated and extrapolated data is available without a proper trend to conclude anything about decline. Further the robust data collected from risk areas habituating various species is especially insufficient to assess any trend especially from arctic and tropical regions of world, for example southern hemisphere [35].

### 1.2 Causes of insect decline

Globally the well-known cause of insect decline is attributed to many factors [36]. The most important among them are habitat destruction due to intensive agricultural practices, urbanization, pesticides use, introduction of new species, climate change and global warming, eutrophication, pollution, genetically engineered plants, UV radiations, ozone depletion and artificial lightening [4] .

In today's agriculture and horticulture crop systems, pesticides and herbicides are used on a large scale, affecting non-target species, insect-plant interactions, soil they live in, and air they breathe.

The excessive applications and quantity of chemical insecticides and herbicides on plants have not adversely influenced the non-target arthropod species but also their host food plants at an alarming rate [37]. Impacts of prevailing climate change and the introduction of exotic species have generated the competition with the indigenous species due to which native species are under pressure, with the consequences the species are probable to succumb to biotic and abiotic factors. The higher $CO_2$ levels in any agro-ecosystem enhance the faster vegetative growth in plants producing higher biomass and lesser nutrients due to reduced photosynthesis [38, 39]. Further, the insect species especially Dipteran (flies) and Dictyopteran (cockroaches) populations may increase however, overall projected insect biomass under higher $CO_2$ levels may decrease ranging from 0.9 to 2.5% per year [40, 41], and insects are losing an average 10–20% of their land every decade, which is horrifying. According to one latest meta-analysis report the intense reductions (up to −80%) in insect abundance and biomass confirmed an observed species richness declines ranging from 20–40% on seasonal basis, especially the decline of dipteran species contributed by consequences of improper functioning of ecosystems [42] due to many factors.

### *1.2.1 Habitat destruction*

Cutting down of trees converting wild land into agriculture, silviculture and other commercial developmental usage can lead to the great decline in the diversity of living organisms all around the world [40, 41] Habitat change is primarily due to human activities and its scope has been expanding over the past centuries because large amount of land has been transformed to provide dwelling, and facilitate transportation and tourism infrastructure, at the expense of natural habitat. Among insect species, order, Coleoptera, Lepidoptera and Hymenoptera are most affected by the habitat destruction [42].

### *1.2.2 Pollution*

Pollution has always been one of the serious threats especially industrial pollution to reduce the insect population and leads to reduction in the viability of insects [37]. Exhaust fumes from cars has increased level of nitrogen dioxide [43], air pollution [44], aquatic pollution, Light pollution [45] are the serious threats to population including insect fauna.Pollution is considered as a major cause for insect decline [42] and environmental pollution Viz., uses of fertilizers and synthetic pesticides in agricultural production, usage of sewage and landfill beaches from urbanized areas, chemicals released from various factories and mining sites. These all cause air pollution accounting to 13% [46]. Toxicity of insecticides are most toxic followed by fungicides and then by herbicide on insect decline. Application of herbicides to any crop land affects more negatively to both terrestrial plants and insect fauna than any other agronomic practices [47]. In rural areas of UK, pesticides found to cause decline in the number of moths and pollinators in Italy [48]. The uses of broad- spectrum insecticide destroy ground developing insects [49] While, the use of systemic insecticides cause reduction in the population of lady bird beetles and butterflies [50]. Besides, nicotinamide and fipronil insecticides showed a very negative impact on aquatic insects [51]. Usage of fipronil results in the reduction in the number of dragonflies [52]. Nicotinamide is considered to be the main cause of reduction of dragon fly population in Japan [53]. Usage of avermectins cause decline in dung beetle population [54].

### *1.2.3 Land use change*

Land usage change is also one of the serious causes for the insect decline. Land use change simply means changing the habitat of many ground dwelling and terrestrial insects, causes their decline and eventually leads to their extinction. Many insects are threatened due to the destruction of various small farms. Small and traditional farms and converting these to more industrialized farms is very detrimental to insect decline [55] so by converting the natural habitat into augmented concrete structure around bunds and other water fills.

### 1.2.4 *Deforestation*

Tropical forests are home to the majority of insects. So by cutting these tropical forests for crops for various purposes is the most serious threat regarding the biodiversity of insects [40]. Cutting down trees on large scale altars rainfall patterns, and insect populations and their developmental activities [45]. Deforestation is directly linked to the decline in the population of insects because when trees are cut the insects dwelling on those trees are ultimately destroyed and considered to be the biggest habitat for living organisms including insects so by cutting the forest causes a serious threat to all the biodiversity including insect faun [42] as forests are considered as stable ecosystem for all living creature.

### 1.2.5 *Agriculture*

In the recent years, agricultural infestation has also accelerated [56] which eventually resulted in the change in the composition of insects. The factors for agricultural infestations are responsible for insect decline. Such as artificial drainage causes the reduction in wetland which eventually causes the habitat loss for many aquatic insects [57], removing Woodland trees eliminating the insect food and shelter structures of insects [58, 59] besides using inorganic micro fertilizers results in eutrophication which eventually affect the aquatic insects like may fly [60] beside this extensive use of pesticides in agricultural practices is also very serious pest population that's the reason that there is it ever growing link of literature linking the insect decline with the agricultural intensification. Agricultural intensification eventually results in homogenization of microhabitat and causes changes in the insect communities. Combination of different waterbodies lead to eutrophication sedimentation in water bodies that causes reduction in various predators [57, 60]. Aquatic plants are very important provide refuge for the insects belonging to the order Odonata [61] and reduction in the insect biodiversity is also caused by the loss of streamflow and river trade water bodies.

### 1.2.6 *Urbanization*

Urbanization is also considered to be one of the main causes for insect decline. Globally, urbanization is increasing day by day which leads to habitat fragmentation and converting the large habitats into smaller areas and converting the forest into agricultural areas and communities [62]. In tropical West Africa, huge decline in beetle and wasp populations were observed due to urbanization [21, 62]. Globally, agricultural fields has been converted into urbanization areas to meet the demands of urban population for their housing and other needs that leads to insect decline where these insects had been living since decades.

### 1.2.7 *Climate change*

Nowadays ecologist and conservationists are working to relate the climatic change with the decline of insects [21]. Climate change has become one of the reasons for the decline of butterfly and wild bees [63]. Insects in tropical regions usually have a very narrow thermal thresholds and very susceptible to temperature fluctuations. Besides, global warming can enhance the population of butterflies, their geographical distribution and more towards northern areas [64]. There are reports that more than half the world's insect populations are declining due to the global warming [65, 66]. Global warming result in a reduction of populations of some dragonflies, bumblebees and stoneflies which are mainly adapted in cold climates and live in the higher altitude [67]. It has a negative impact on some pollinator and beetles which are located in Mediterranean region., it might increase extinction of many mountain species [68]. The clear evidences of climate change leading to the reduction or decline in the biomass of arthropods in the rainforests of Caribbean Islands [69].

### *1.2.8 Invasive species*

Introduction of non-native or invasive species occur in a particular ecosystem is a threat on the existing population. In areas of human occupation and introduction of invasive plants reduce the insect herbivore loads more than 90%. The reason behind the collapse of honeybee colony in various countries is the introduction of various exotic parasites and pathogens [70] and also leads to the decline of wild bees in North America [71]. The spread of vartmaan destructor mite (*Aethina tumida*) is threat to apiculture industry [72]. There are several reports regarding the impact of invasive plant and animal species on the native insect species.

### *1.2.9 Pesticides*

The use of pesticides is the main causes for the insect decline. Aerial application of pesticides is directly linked with the loss of flying insects and population of pollinators. Imidacloprid and Thiacloprid insecticides have a negative effect on bee navigation but widely used for the protection of crops [73], Nicotinoids are the main cause for the decline of dragonflies in Japan [74] and avermectins for the dung beetles in many countries [75].

### *1.2.10 Roads, railway networks and air ways*

Transport infrastructures such as roads and railways and airways are very important nowadays for the basic necessities for human population [76]. Despite being extremely help for the human civilization but very dangerous for insect biodiversity, fragmentation of land due to the construction of roads, railways which causes habitat loss of many insect species [76] found higher mortality rates at intermediate traffic volume compared to high and low traffic rates. Besides, insect crossing on the roads, collusion of insect with vehicles and death of soil borne insects during road constructions [77].

### *1.2.11 Effects of insect decline*

Insects are integral part of ecosystem functioning. The decline of insect populations has direct impact on ecosystems, animal populations, plants, herbs, shrubs and in end on human begins [78]. The structural and functional base of the ecosystems are made by insects and as per global review, the decline of insect populations if not managed and mitigated would have disastrous and cataclysmal effects on global ecosystem. The parasitoids and predators (birds and mammals) which directly feed or host on insects are affected by insect decline [79]. The decline of bees and beneficial bugs reduce the pollination of diverse plant populations and the biological waste disposal [4]. The reports of zoological survey of London revealed that insect decline corresponds to losses of instrumental values and the species intrinsic values.

### *1.2.12 Evidence of insect decline*

Broadly speaking there are three most important metrics capturing and reporting the insect decline, as an evidence for determination of insect decline globally [80]. The first important component is insect population abundance which determines the numerical total of individuals in any particular ecosystem. The insect abundance is measured differently under different contexts; however, the overall intend refers to number of insect species per meter square of plant, in any assembly, per unit geographical area or sum total of insects present globally. Biomass is second important metrics of insect estimation through their total weight irrespective of insect species into consideration. Biodiversity is the third and important metrics [81] based on broader scale of measurements and gives a metadata about insect species existing globally. Like abundance the term insect biodiversity is used under different contexts and the reduction in biodiversity presents an alarming threat. Those species that have vanished locally and some particular species has gone entirely extinct form earth.

The available literature says that most of the studies concerning global insect decline base their studies on few metrics [82] either abundance or biomass or biodiversity or combination of all three and very few meta-data analysis studies recorded all the three metrics for proper estimation of insect decline. The available data of direct evidence of diversity decline and loss is scattered and inadequate for all the three metrics and it becomes difficult to comprehend the true figures of total insect reduction globally. Therefore, the estimates of insect diversity loss at cosmogonic level are inclined to involve the generalization and extrapolating from existing abundance or/and biomass data; however, the true worldwide extinctions figures are demanding to discern and determine.

David Wagner reviewed the literature and proposed that presently the Holocene extinction contributing the species loss to about 100–1000 times as compared to previous pace. Some authors proposed even faster extinction rate and collaborated with Wagner's opinion that rapid decline in insect abundance would have a serious ecological impact. The global decline of megafauna is related to human activities not to climate change [83].

### *1.2.13 Relationship between metrics based on historical inferences*

Ecologists provided different views of relationships among three important insect ecological metrics [84]. Few hypothesized an independent relationship while others explained dependence of metrics on each other either directly or indirectly. For example, reduction in biomass might not necessarily involve a decrease in abundance or diversity. The reduction of particular species means shrinking of biodiversity and consequently the reduction of insect abundance and finally the species is getting smaller in range of expansion and the area is becoming lesser in richness. In real, abundance and biomass are closely related and both sowing decline depending on various biotic and abiotic factors. Therefore, insect diversity is often, though not related to three metrics. The Rothamsted Research Institute from UK conducted an insect survey using suction traps during 1964 and compiled a most standardized long-term data on insects in the world [14, 85]. Suction tarps were installed effectively positioning upside-down Hoovers running 24/7 sampled the air for all migrating insects. Revaluating the data during 2000–2017, the James Bell in an interview in 2017 announced that insect populations in Scotland has reduced while as figures from England has remained comparatively stabilized. The review from [86] and other reports that "Of the all insect populations with IUCN- documented population trend, 33% are declining and 30–60% of species per order are in declining ranges. With regards to insect pollinators the populations are declining globally both in abundance and diversity; the loss is contributed by human-caused disturbance of vertebrates and invertebrates as the consequence of Anthropocene defaunation [87].

The reports of higher decline in insect biomass were recorded by Krefeld Entomological using malaise traps at Germany. For this study a total of 63 locations were chosen comprised of 57 from Nordrhein-Westfalen, one each from Rheinland-Pfalz and Brandenburg and 4 other from nearby areas. Studies concluded that there were a seasonal and mid-summer biomass decline of 76% and 82% in flying insects. The decline were contributed by several factors especially change in weather, land use and habitat features. The flying insects especially butterflies, moths and wild bees were major contributors to total decline proportion and based on the results of Krefeld German government have established an Action Programme for the Insect Protection.

The decline in arthropods were reported from Puerto Rico based on the surveys and measurements during 1976 and 2012 [69]. During 36 years of studies, the biomass losses evaluated were 98% and 78% for ground foraging and canopy dwelling arthropods with annual losses of 2.7 and 2.2% respectively. The reasons behind this rate of fast decline were average high temperatures; since the arthropods from tropical areas do not tolerate high temperatures. In tropics rapid cold-hardening, ice-interface desiccation and the daily resetting of critical thermal thresholds affecting mortality and mobility with temperature as the most important factor having high influence on insect physiological processes to determine ecological outcomes and survival under harsh conditions [88].

Estimated declines of 84% in butterflies were observed from Netherland during 1890 to 2017 [66]. Further Swiss Academy of Natural Sciences recorded an estimated decline of about 60% in insect- eating birds urging the authorities for the immediate action to resolve the causes of insect decline. Similarly, in another studies of 2019, published in Biological Conservation showed a decline in insect populations from US and Western Europe. Further, the studies showed an annual loss of 2.7% in biomass; and it was hypothesized that this rate of decline may lead to mass extinction of 40% species over next 45–55 years. Extinction of insects groups like butterflies, moths, bees, dragonflies, beetles, dipteran flies, and Orthopteran (grasshoppers, crickets) and Hemiptera (aphids) are more susceptible for loss in terms of abundance, diversity and total biomass. According to global assessment report (2019) regarding various important metrics of insects in environment, the global trends in insect populations were not clearly determined but rapid decline have been well documented in many hot spot areas [89]. Local declines in insect densities (bees and butterflies) and overall global decline in abundance have been observed. Some rapid decline was contributed by many factors like large scale land use changes, rising temperatures, and reduction in conducive habitats. The reports from a meta-analysis studies published in Science showed an average abundance decline of 9% in terrestrial insects; contrary to this an average 12% increase in freshwater abundance were also reported [9]. Rise in aquatic insect populations have been contributed by many important factors such as the sanitary measures and other rapid actions taken by governments in present climate change era, the freshwater bodies have remained clean and people residing near these natural water bodies have been provided awareness for the preservation, conservation and maintenance of the hygienic environment that overall boasted the aquatic insect population besides reducing the water pollinations. Some other studies at US have also presented similar data from different ecosystems with both decline and increase reports showing that overall insect abundance have changed but no net change in biomass has occurred.

Depending on the specific region, the butterfly populations across a large part of North America are declining (North America), increasing (South East part) or stable based on area sampled [90]. The reasons for these irregular variations were climate change, especially the emission of greenhouse gases affecting butterflies and moths.

### 1.3 Insect conservation

The insect decline is a global issue and much of the efforts to retain and restore biodiversity at national or international levels are addressed to US as component of Convention on Biological Diversity [91]. The communications and reports typically describe policies and planning to save further loss of diversity through conservation and restoration of habitats, host plants, and measures to reduce disturbances to protect the particular threatened taxa [92]. For conservation point, the prime importance is given to insect pollinators as being most essential and integral part of crop production and the global efforts to reduce their decline are at high priority especially focusing on conservation of bumble bees [93], honey bees and some other solitary bees. The German Environment Ministry started an action Programme for insect protection which aims at promoting insect habitat in diverse agricultural landscapes by reducing pesticides use, light pollination and pollutants in soil and water. The United Nations initiated a compressive sustainable development goals by drafting a policy making community transition from perceiving insects as enemies and injurious to providers of ecosystem services. Entomological Society of America advising the farmers to maintain plant diversity in their farms by leaving some buffer areas, natural habitats, leaf litter and dead woods for insect proliferation and breeding. Similarly, the Xeres Society US stressed collaborators to promote invertebrate conservation for applied research, and advocacy and to promote public outreach and education. The project aimed at the rehabilitation of natural habitat for endangered species, conservation of insect pollinators, restoration and their protection. Further, the phone apps like iNaturalist for photography and identification of specimens and the programs such as City Nature Challenge, National Moth Week and Monarch Butterfly Conservation were initiated.

### 1.4 Global decline of insect studies

Awareness about insect significance to environment is of high value [94]. Therefore, the lack of this awareness is contributed to the global decline of studies of entomology and taxonomy. The mention in Entomology Congress in US stated that the studies of entomology are themselves as an endangered species and according to one survey the world has lost nearly all experts. General biology courses in colleges and universities have given less attention to insect science and number of specialists are decreasing. Further, the studies related to decline trends, management, diversity and other metrics estimation involved collecting, killing, trapping and ward off which have some ethical issues for conservationists [15].

#### *1.4.1 Biodiversity loss*

Biodiversity loss comprised of devastation, extermination and extinction of insect species worldwide [95]. The disappearance of organisms from different natural reservoirs and ecosystems results in a temporary or irreversible loss of insect biodiversity, depending on whether various disruption factors are reversible (ecological restoration and resilience) or effectively permanent (ecological restoration and resilience) (land loss, erosion, deforestation). Human behaviors of different types beyond reasonable range are causing the most permanent changes in the twenty-first century, and leading to high biodiversity losses. The recent and the big irreversible global species loss is highly catastrophic and tragic phenomenon compared to regional loss in species composition. The regional and minor species composition changes from stable state have a huge negative impact on food web and food chain [96]. Since decreasing and extinction of one species would adversely impact the entire cycle of food chain.

The disturbances and breaks in food chain ultimately lead to diminishing biodiversity, unstabilization of ecosystem services. Decline in insect biodiversity simultaneously presents an immediate threat to food security [97] and moreover, have a permanent and adverse effects on health and wealth of humans. Since years, the International Environment Organization has led a campaign to avoid biodiversity loss by combining public health and biodiversity conservation into a single health solution that can be used as part of international policy. The UN Convention on Biological Diversity aimed at restoration and conservation of biodiversity (2021) loss and proactive measures; Sustainable Development Goal 15 for "Life on Land" and Sustainable Development Goal 14 for "Life Below Water", and the UN Environment Programme were designed to focus for Making Peace with Nature. All these efforts remained either unimplemented or fail to meet their targets on real grounds.

### *1.4.2 Holocene extinction*

Since centuries human activities are pushing environment beyond the recovery and revival to the ultimate catastrophic events of extinction [21, 98]. According to historical prospective of species extinction, the Holocene extinction is considered as the 6$^{th}$major mass extinction event also referred to as anthropocene extinction. The anthropocene extinction is an ongoing loss of species during the current Holocene epoch with the consequences of various human activities especially started with the onset of technological revolution. The diverse spread of destruction of species from biologically diverse habitats is viewed to be unknown and unrecognized. Most of the extinct insect species were either not known to science or yet undiscovered without knowing their cause of extinction. The present speed of species extinction is calculated to be as fast as 100 to 1000 times higher than naturally occurring previously recorded extinction rates [98]. The species at megafauna disappearance during final phase of last glacial period were known to be highly sensitive to predation died shortly at the beginning of hunting activities by human across the earth especially from Africian region at an onset of Holocene era of extinction event near Pleistocene–Holocene boundary frequently known as quaternary extinction event [99].

### *1.4.3 Defaunation*

The functional extinction of insect species [99] at global, local and regional level is referred as defaunation from ecological communities; triggered by growth and spread of human populations coupled with the advancement of latest technologies; ultimately leading to an unlimited and unbearable exploitation of the ecosystem in which diverse insect species are living. The diminishing and dwindling of the invertebrate species from ecological communities result into the empty forests resulting into species disappearance and reduced abundance. The estimates of more than 50% of all wild life species were lost due to defaunation in last 40 years of Holocene era [100]. The surprising example is from year 2016 contributed by 68% species defaunation in terms of disappearance and reduced abundance compared to even higher species loss of 70% from South America [101]. Regarding this fast defaunation in current era the global gathering of 15000 scientists during 2017 called for a second warning to humanity for development and implementation of stringent policies to mitigate the Defaunation and exploitation of the natural reservoirs to ensure the safeguarding of remained threatened species.

The endangered insect species are designated to be facing a high threat of extinction viewed by Zoological survey of India (ZSI) in the natural ecosystems. Therefore in 2017 the US International Union for Conservation of Nature (IUCN) sort out a total of 343 endangered insect species comprised of total 5.7% as endangered. Further, the IUCN also highlighted 21 more subspecies as endangered for extinction; however no subpopulations were evaluated by IUCN. For IUCN to consider a species as endangered must fulfill some prerequisites to classify the taxa facing a very high risk of extinction in near future. The critically high endangered species are known to be nearly 538; however, 1702 species (28% of total evaluated) were considered as data deficient with minimum available information to full assessment and for determining their conservation status [102]. However, IUCN noted them as with the same degree of attention as threatened taxa till their status can be assessed.

### 1.5 Mitigation of insect decline

To maintain the ecological balance of every existing ecosystem for survival is essential. Every creature from microscopic to macroscopic in structure has its own importance and credibility to maintain the existence of the living world. All terrestrial insects provide resources for higher trophic levels, especially for many vertebrates [103]. Natural pest regulation relies on insect presence and their trophic interactions [104]. The local extinction of plants relying on pollinators is related to diminishing of pollinators [105]. The annual value of pollination for agriculture plants is estimated to be 200–600 billion US$ [106]. To maintain proper functionality and to mitigate the adverse effects of natural calamities of the environment. Diverse species of insects are under natural as well anthropogenic risks which acts as the main drivers to decline the insect population. Literature survey revealed that the insect population is declining and at an alarming pace at the present scenario. Therefore, research work in this field is not considered important by the entomologists but needs to provide a very special attention. Some of the basic and important features are presented that plays an important role to mitigate the population decline of insects.

#### *1.5.1 Conservation of high quality habitat*

High-quality habitat protection and fortification must be given first significance. Various rules and regulations have been framed by various national as well as international organization to safeguard and protect the valuable habitats of insects. Laws framed to protect habitats and other natural surroundings to prevent any kind of destruction or devastation at the European Union (EU) Level, and must be stringently executed in all the member states without any latency. Even though to safeguard diverse species and different ecosystems by the enactment of the Natura 2000 network (N2000), and 18% surface of land is covered by European Union, it is the Habitats Decree (directive) that emphases on susceptible species. Though, many of the targeted species enumerated on the annexes of the Habitats Decree Directive characterize peripheral or even relict populations, and thus the EU often does not cover the primary of their global circulation range [107]. Moreover, Birds and plants were primarily targeted in the Habitats Directive but now there is urge that invertebrates are be included [108]. To meet the demands of new species or to achieve conservation goals, more focus should be given to develop or reinstated high quality habitats in the farming environment, as the existing environment, as per the Habitats directive, would not be able to assure the survival of species of insects [109]. So a new suggestion is presented that public interest must have precedence over individual proprietorship rights.

## 1.6 Increasing landscape permeability

"A healthy environment is a key to a healthy living organism", means that if a living creature is nurtured in a healthy environment, that depicts the growth of an individuals with enhanced capability traits. Healthy environment functional networks fosters/helps to long survival of many species [110]. A conclusion is drawn that special economic packages must be supplied to those organizations or agronomists who can employ ecofriendly practices. To increase landscape permeability acts an important factor to mitigate the population decline of insects, so special methods and techniques must be encouraged by policy makers as well as government organizations.

Landscape permeability refers as "that area of habitat of an individual which provides free passage without any obstacle to fulfill survival requirements "Land connectivity can be increased by extending field margins as well as roadside extensions, which in turn increase more area for flowering plants that ultimately leads to insect friendly conditions and thus helps in mitigating insect decline [111]. Beetle movement to adjacent areas is fortified by grassy fields or grassy strips [112]. While, barren lands acts a favorable and healthy environment for bumble bees and butterflies in Finland [113]. Such kinds of existing strategies has been proved economical but also easy to implement and thus easy to understand the insect decline.

### *1.6.1 Safeguarding habitat quality*

Chemical fertilizers and their other insolvable constituents are considered one of the ultimate threats to the population of different species of insects. Intensive farming is excessively dependent upon the chemical fertilizers to yield better output from the crops. The applications of these fertilizers would be reduced to an acceptable level by using alternative organic fertilizers. On the basis of various surveying methods, a large number of food chains and webs are disturbed or completely damaged by the excessive use of these toxic chemicals. The negative impacts have developed so many detrimental impacts and needs to be reduced both in agricultural and urban areas. Neonicotinoids, a kind of pesticides, has reduced bees' population to large extent [114]. Thus there must be special and regular conferences to make awareness among the masses regarding the damaging and toxicant effects of these products of the fertilizers. There are many existing alternative methods to adopt and to implement in place of chemical fertilizers and thus helps to mitigate the declining population of insects to large extent. Organic farming plays an imperative role to sustain biodiversity and enhance insect population with a balanced manner [111, 115, 116]. However, there is still requirement to improve this research further to improve yields by use of organic farming [117].

### *1.6.2 Mitigate soap run-off from washing*

Soap consists of long chain fatty acids which are harmful and detrimental to insects. Soaps are used to wash motorbikes, cars, buildings or washing clothes accompanied by harmful pollutants such as heavy metals, ammonia etc. that are redirected into different water bodies [118]. There are diverse types of aquatic insects heavy affected by such chemicals i.e. they inhibit their metabolic system and thus heavily affects their normal population growth. The wise methods needs to be taken to avoid the soap run-off water from washing to come in direct contact with existing water bodies and just to reduce the decline rate of aquatic insects.

### 1.7 Limit use of artificial light at night (ALAN)

Light pollution during night time has adversely affected the insect population and is increasing at an alarming pace, also in various bio diverse areas it has shown two fold growth [119]. These beautiful and wonderful creatures are attracted to light and thus fall prey to artificial lights either by exhaustion or predation [120]. Analytical work has depicted that nocturnal moth's shows downhill growth in Europe against day flying insects [121]. Reproductive process in fire flies has also drastically affected by the exposure of artificial lights [122]. Artificial lights have completely dominated the life of a common man and the visuals seen on roads, in parks, in malls, cars or other light poles acts as traps for insects.

No doubt such designs and decorations are the foremost need of the modern life but the negative implications of artificial lights upon insects must be kept in consideration. To mitigate the declined insect population, awareness though training camps, social media or via other government agencies should be encouraged among people, to use minimum possible artificial lights or either to stop unnecessary light systems. Moreover government at state level or at national must join hands with light designing experts to discover or design insect friendly light systems to safeguard the future generations of insects where trapping of insects are not needed.

#### *1.7.1 Mass trapping and mating disruption of insects*

Mass trapping and mating disruption using pheromone technologies are the latest techniques for pest management [123] as they are species specific technologies as there will be no effect on other close associates in the environment. Besides, no pollution in the environment, no pesticide residues in fruits, secondary pest outbreaks. Though the predators and parasitoids which living and feeding on these target insect pests which are being mass trapped or suppressed in the environment due to mating disruption indirectly affecting the ecological balance and various tritrophic levels.

## 2. Conclusion

Ecological balance refers when all natural existing system work harmoniously with each other and without having any negative consequences. The completion of these natural systems is achieved only if organisms from microscopic level to giant organisms are taken a good care for their survival. However, the importance and prominence of insects, which falls under kingdom arthropods, are either neglected or unaware by a common man. But to maintain the continuity of life on earth, the role of insects is essential especially the pollination of various plant species. Decline in the growth of insect population has become a societal, scientific, economic challenge for entomologist and researchers. Thus special philosophical, political, scientific, and psychological measures are need of the hour to mitigate the decline of insect population for human survival especially the pollinating, parasitizing and predating insects. Therefore, various methods are presented and defined which would help to mitigate declining population of insects with acceptable levels and to safeguard the rich and healthier future of insects and for the human race.

## Author details

Showket A. Dar[1], Mohmmad Javed Ansari[2], Yahya Al Naggar[3], Shafia Hassan[4], Syed Nighat[4], Syed Burjes Zehra[5], Rizwan Rashid[6], Mudasir Hassan[7] and Barkat Hussain[8]*

1 Divsion of Entomology, Krishi Vigyan Kendra Zanskar-194302, Sher-e-Kashmir University of Agricultural Sciences and Technology of Kashmir, Srinagar, J&K, India

2 Department of Botany, Hindu College Moradabad (Mahatma Jyotiba Phule Rohikhand University Bareilly), Uttar Pradesh, India

3 Zoology Department, Faculty of Science, Tanta University, Tanta, Egypt

4 Department of Zoology, University of Kashmir, J&K, India

5 Divsion of Vegetable Science, MAR & ES, Kargil, Sher-e-Kashmir University of Agricultural Sciences and Technology of Kashmir, Srinagar, J&K, India

6 Divsion of Vegetable Sciences, Sher-e-Kashmir University of Agricultural Sciences and Technology of Kashmir, Shalimar, Srinagar, J&K, India

7 Department of Agriculture, J&K, India

8 Divsion of Entomology, Sher-e-Kashmir University of Agricultural Sciences and Technology of Kashmir, Srinagar, J&K, India

*Address all correspondence to: bhatbari@rediffmail.com

## References

[1] Wagner DL, Grames EM, Forister ML, Berenbaum MR, Stopak D. Insect decline in the Anthropocene: Death by a thousand cuts. Proceedings of the National Academy of Sciences. 2021 Jan 12;118(2).

[2] Ceballos G, Ehrlich PR, Raven PH. Vertebrates on the brink as indicators of biological annihilation and the sixth mass extinction. Proceedings of the National Academy of Sciences. 2020 Jun 16;117(24):13596-13602.

[3] Sánchez-Bayo F, Wyckhuys KA. Worldwide decline of the entomofauna: A review of its drivers. Biological conservation. 2019 Apr 1;232:8-27

[4] Diamond JM. The present, past and future of human-caused extinctions. Philosophical transactions of the Royal Society of London. B, Biological Sciences. 1989 Nov 6; 325(1228): 469-477.

[5] Monbiot G. Insectageddon: Farming is more catastrophic than climate breakdown. The Guardian. 2017 Oct 20;20.

[6] Wilson EO. What is nature worth?. The Wilson Quarterly (1976-). 2002 Jan 1;26(1):20-39.

[7] Hallmann CA, Sorg M, Jongejans E, Siepel H, Hofland N, Schwan H, Stenmans W, Müller A, Sumser H, Hörren T, Goulson D. More than 75 percent decline over 27 years in total flying insect biomass in protected areas. PloS one. 2017 Oct 18;12(10): e0185809.

[8] Hallmann CA, Zeegers T, van Klink R, Vermeulen R, van Wielink P, Spijkers H, van Deijk J, van Steenis W, Jongejans E. Declining abundance of beetles, moths and caddisflies in the Netherlands. Insect Conservation and Diversity. 2020 Mar;13(2):127-139.

[9] Van Klink R, Bowler DE, Gongalsky KB, Swengel AB, Gentile A, Chase JM. Meta-analysis reveals declines in terrestrial but increases in freshwater insect abundances. Science. 2020 Apr 24;368(6489):417-420.

[10] McDermott A. News feature: To understand the plight of insects, entomologists look to the past. Proceedings of the National Academy of Sciences. 2021 Jan 12;118(2).

[11] Manning DW, Sullivan SM. Conservation across aquatic-terrestrial boundaries: Linking continental-scale water quality to emergent aquatic insects and declining aerial insectivorous birds. Frontiers in Ecology and Evolution. 2021 Feb 26;9:68.

[12] Manning DW, Sullivan SM. Conservation across aquatic-terrestrial boundaries: Linking continental-scale water quality to emergent aquatic insects and declining aerial insectivorous birds. Frontiers in Ecology and Evolution. 2021 Feb 26;9:68.

[13] Mingarro M, Cancela JP, BurÓn-Ugarte A, García-Barros E, Munguira ML, Romo H, Wilson RJ. Butterfly communities track climatic variation over space but not time in the Iberian Peninsula. Insect Conservation and Diversity. 2021.

[14] Bowler D, Eichenberg D, Conze KJ, Suhling F, Baumann K, Boensel A, Bittner T, Drews A, Guenther A, Isaac N, Petzold F. Winners and losers over 35 years of dragonfly and damselfly distributional change in Germany. BioRxiv. 2020 Jan 1.

[15] Pilotto F, Kühn I, Adrian R, Alber R, Alignier A, Andrews C, Bäck J, Barbaro L, Beaumont D, Beenaerts N, Benham S. Meta-analysis of multidecadal biodiversity trends in Europe. Nature communications. 2020 Jul 13;11(1):1-1.

[16] Braby MF, Yeates DK, Taylor GS. Population declines and the conservation of insects and other terrestrial invertebrates in Australia. Austral Entomology. 2021 Feb; 60(1):3-8

[17] Hallmann CA, Ssymank A, Sorg M, de Kroon H, Jongejans E. Insect biomass decline scaled to species diversity: General patterns derived from a hoverfly community. Proceedings of the National Academy of Sciences. 2021;118(2).

[18] Harvey JA, Heinen R, Armbrecht I, Basset Y, Baxter-Gilbert JH, Bezemer TM, Böhm M, Bommarco R, Borges PA, Cardoso P, Clausnitzer V. International scientists formulate a roadmap for insect conservation and recovery. Nature Ecology & Evolution. 2020 Feb;4(2):174-176.

[19] Habel JC, Teucher M, Gros P, Schmitt T, Ulrich W. Land use and climate change affects butterfly diversity across northern Austria. Landscape Ecology. 2021 Apr 22:1-4.

[20] Hopkins TL. Extinction of the rocky mountain locust. BioScience. 2005 Jan 1;55(1):80-82.

[21] Cardoso P, Barton PS, Birkhofer K, Chichorro F, Deacon C, Fartmann T, Fukushima CS, Gaigher R, Habel JC, Hallmann CA, Hill MJ. Scientists' warning to humanity on insect extinctions. Biological Conservation. 2020 Feb 1;242:108426.

[22] Samways MJ, Barton PS, Birkhofer K, Chichorro F, Deacon C, Fartmann T, Fukushima CS, Gaigher R, Habel JC, Hallmann CA, Hill MJ. Solutions for humanity on how to conserve insects. Biological Conservation. 2020 Feb 1;242:108427.

[23] Wilcove DS. No Way Home: The Decline of the world's Great Animal Migrations. Island Press; 2010 Aug 30.

[24] Penney D, Jepson JE. Fossil Insects: An Introduction to Palaeoentomology. Siri Scientific Press; 2014 Jul 31.

[25] Keller G, Armstrong H, Courtillot V, Harper D, Joachimski M, Kerr A, MacLeod N, Napier W, Palfy J, Wignall P. Volcanism, Impacts and Mass Extinctions (Long Version). Geological society of London, London. 2012.

[26] Labandeira CC, Sepkoski JJ. Insect diversity in the fossil record. Science. 1993 Jul 16;261(5119):310-315.

[27] Eriksson O. The importance of traditional agricultural landscapes for preventing species extinctions. Biodiversity and Conservation. 2021 Mar 1:1-7.

[28] Fukano Y, Soga M, Fukuda M, Takahashi Y, Koyama M, Arakawa Y, Miyano N, Akiba Y, Horiguchi M. Debut of an endangered bird in zoos raises public interest, awareness and conservation knowledge of the species. Animal Conservation.2021. https://doi.org/10.1111/acv.12693:

[29] Stork NE, Coddington JA, Colwell RK, Chazdon RL, Dick CW, Peres CA, Sloan S, Willis K. Vulnerability and resilience of tropical forest species to land-use change. Conservation Biology. 2009 Dec;23(6):1438-1447.

[30] Welti EA, Zajicek P, Ayasse M, Bornholdt T, Buse J, Dziock F, Engelmann RA, Englmeier J, Fellendorf M, Forschler MI, Frenzel M. Climate, latitude, and land cover predict flying insect biomass across a German malaise trap network. bioRxiv. 2021 Jan 1.

[31] Amendt J. Insect decline—A forensic issue?. Insects. 2021 Apr;12(4):324.

[32] Dunn RR. Modern insect extinctions, the neglected majority.

Conservation biology. 2005 Aug;19(4):1030-1036.

[33] Almond RE, Grooten M, Peterson T. Living Planet Report 2020-Bending the Curve of Biodiversity Loss. World Wildlife Fund; 2020.

[34] Staudinger MD, Grimm NB, Staudt A, Carter SL, Chapin FS. Impacts of Climate Change on Biodiversity, Ecosystems, and Ecosystem Services. United States Global Change Research Program, Washington, DC. 2012 Jul.

[35] Balmori A. Electromagnetic radiation as an emerging driver factor for the decline of insects. Science of The Total Environment. 2021 Jan 28:144913.

[36] Brühl CA, Zaller JG. Biodiversity decline as a consequence of an inappropriate environmental risk assessment of pesticides. Frontiers in Environmental Science. 2019 Oct 31;7:177.

[37] Tudi M, Daniel Ruan H, Wang L, Lyu J, Sadler R, Connell D, Chu C, Phung DT. Agriculture development, pesticide application and its impact on the environment. International Journal of Environmental Research and Public Health. 2021 Jan;18(3):1112.

[38] Branscome DD, Koehler PG, Oi FM. Influence of carbon dioxide gas on German cockroach (Dictyoptera: Blattellidae) knockdown, recovery, movement and feeding. Physiological Entomology. 2005 Jun;30(2):144-150.

[39] Bhargava S, Mitra S. Elevated atmospheric CO2 and the future of crop plants. Plant Breeding. 2021 Feb;140(1):1-1.

[40] Maxwell SL, Fuller RA, Brooks TM, Watson JE. Biodiversity: The ravages of guns, nets and bulldozers. Nature News. 2016 Aug 11;536(7615):143.

[41] Fischer M, Rounsevell M, Rando AT, Mader A, Church A, Elbakidze M, Elias V, Hahn T, Harrison PA, Hauck J, Sandstrom C. The Regional Assessment Report on Biodiversity and Ecosystem Services for Europe and Central Asia: Summary for Policymakers. IPBES secretariat; 2018.

[42] Whittaker JB. Presidential address: Insects and plants in a changing atmosphere. Journal of Ecology. 2001 Aug 1;89(4):507-518.

[43] Campbell SA, Vallano DM. Plant defences mediate interactions between herbivory and the direct foliar uptake of atmospheric reactive nitrogen. Nature communications. 2018 Nov 9;9(1):1-7.

[44] Hain FP. Interactions of insects, trees and air pollutants. Tree Physiology. 1987 Mar 1;3(1):93-102.

[45] Arimoro FO, Ikomi RB. Ecological integrity of upper Warri River, Niger Delta using aquatic insects as bioindicators. Ecological indicators. 2009 May 1;9(3):455-461

[46] Dudley N, Alexander S. Agriculture and biodiversity: A review. Biodiversity. 2017 Jul 3;18(2-3):45-49.

[47] Hyvönen T, Salonen J. Weed species diversity and community composition in cropping practices at two intensity levels–a six-year experiment. Plant ecology. 2002 Mar;159(1):73-81.

[48] Brittain CA, Vighi M, Bommarco R, Settele J, Potts SG. Impacts of a pesticide on pollinator species richness at different spatial scales. Basic and Applied Ecology. 2010 Mar 1;11(2):106-115.

[49] Lundgren P, Kiryukhin A, Milillo P, Samsonov S. Dike model for the 2012-2013 Tolbachik eruption constrained by satellite radar interferometry observations. Journal of Volcanology and Geothermal Research. 2015 Dec 1;307:79-88.

[50] Krischik V, Rogers M, Gupta G, Varshney A. Soil-applied imidacloprid translocates to ornamental flowers and reduces survival of adult Coleomegilla maculata, Harmonia axyridis, and Hippodamia convergens lady beetles, and larval Danaus plexippus and Vanessa cardui butterflies. PloS one. 2015 Mar 23;10(3):e0119133.

[51] Beketov MA, Liess M. Potential of 11 pesticides to initiate downstream drift of stream macroinvertebrates. Archives of environmental contamination and toxicology. 2008 Aug;55(2):247-253.

[52] Jinguji H, Thuyet DQ, Uéda T, Watanabe H. Effect of imidacloprid and fipronil pesticide application on Sympetrum infuscatum (Libellulidae: Odonata) larvae and adults. Paddy and Water Environment. 2013 Jan;11(1):277-284.

[53] Beketov MA, Schäfer RB, Marwitz A, Paschke A, Liess M. Long-term stream invertebrate community alterations induced by the insecticide thiacloprid: Effect concentrations and recovery dynamics. Science of the Total Environment. 2008 Nov 1;405(1-3): 96-108.

[54] Wardhaugh KG, Mahon RJ. Avermectin* residues in sheep and cattle dung and their effects on dung-beetle (Coleoptera: Scarabaeidae) colonization and dung burial. Bulletin of Entomological Research. 1991 Sep;81(3):333-339.

[55] Scholes RJ, Montanarella L, Brainich E, Barger N, Ten Brink B, Cantele M, Erasmus B, Fisher J, Gardner T, Holland TG, Kohler F. IPBES (2018): Summary for policymakers of the assessment report on land degradation and restoration of the Intergovernmental Science-Policy Platform on Biodiversity and Ecosystem Services.

[56] Attwood SJ, Maron M, House AP, Zammit C. Do arthropod assemblages display globally consistent responses to intensified agricultural land use and management?. Global Ecology and Biogeography. 2008 Sep;17(5):585-599.

[57] Grab H, Branstetter MG, Amon N, Urban-Mead KR, Park MG, Gibbs J, Blitzer EJ, Poveda K, Loeb G, Danforth BN. Agriculturally dominated landscapes reduce bee phylogenetic diversity and pollination services. Science. 2019 Jan 18;363(6424):282-284.

[58] Hain FP. Interactions of insects, trees and air pollutants. Tree Physiology. 1987 Mar 1;3(1):93-102.

[59] Habel JC, Samways MJ, Schmitt T. Mitigating the precipitous decline of terrestrial European insects: Requirements for a new strategy. Biodiversity and Conservation. 2019 May;28(6):1343-1360.

[60] Burdon FJ, McIntosh AR, Harding JS. Habitat loss drives threshold response of benthic invertebrate communities to deposited sediment in agricultural streams. Ecological Applications. 2013 Jul;23(5): 1036-1047.

[61] Roy HE, Brown PM. Ten years of invasion: Harmonia axyridis (Pallas) (Coleoptera: Coccinellidae) in Britain. Ecological Entomology. 2015 Aug;40(4):336-348.

[62] Guenat S, Kunin WE, Dougill AJ, Dallimer M. Effects of urbanisation and management practices on pollinators in tropical Africa. Journal of Applied Ecology. 2019 Jan;56(1):214-224.

[63] Bartomeus I, Ascher JS, Wagner D, Danforth BN, Colla S, Kornbluth S, Winfree R. Climate-associated phenological advances in bee pollinators and bee-pollinated plants. Proceedings of the National Academy of Sciences. 2011 Dec 20;108(51):20645-20649.

[64] Hussain B, War AR, Pfeiffer DG. Mapping foliage damage index and monitoring of Pieris brassicae by using GIS-GPS technology in cole crops. Journal of Entomology and Zoology Studies. 2018; 6:933-938.

[65] War AR, Taggar GK, War MY, Hussain B. Impact of climate change on insect pests, plant chemical ecology, tritrophic interactions and food production. International Journal of Clinical and Biological Sciences. 2016;1(02):16-29.

[66] Warren MS, Maes D, van Swaay CA, Goffart P, Van Dyck H, Bourn NA, Wynhoff I, Hoare D, Ellis S. The decline of butterflies in Europe: Problems, significance, and possible solutions. Proceedings of the National Academy of Sciences. 2021;118(2).

[67] Ball-Damerow JE, M'Gonigle LK, Resh VH. Changes in occurrence, richness, and biological traits of dragonflies and damselflies (Odonata) in California and Nevada over the past century. Biodiversity and Conservation. 2014 Jul;23(8):2107-2126.

[68] Menéndez R, Megías AG, Hill JK, Braschler B, Willis SG, Collingham Y, Fox R, Roy DB, Thomas CD. Species richness changes lag behind climate change. Proceedings of the Royal Society B: Biological Sciences. 2006 Jun 22;273(1593):1465-1470.

[69] Lister BC, Garcia A. Climate-driven declines in arthropod abundance restructure a rainforest food web. Proceedings of the National Academy of Sciences. 2018 Oct 30;115(44): E10397-E10406.

[70] Goulson D, Nicholls E, Botías C, Rotheray EL. Bee declines driven by combined stress from parasites, pesticides, and lack of flowers. Science. 2015 Mar 27;347(6229).

[71] Williams PH, Osborne JL. Bumblebee vulnerability and conservation world-wide. Apidologie. 2009 May 1;40(3):367-387.

[72] VanEngelsdorp D, Caron D, Hayes J, Underwood R, Henson M, Rennich K, Spleen A, Andree M, Snyder R, Lee K, Roccasecca K. A national survey of managed honey bee 2010-11 winter colony losses in the USA: Results from the bee informed partnership. Journal of Apicultural Research. 2012 Jan 1;51(1):115-124

[73] van Lexmond MB, Bonmatin JM, Goulson D, Noome DA. Worldwide integrated assessment of the impact of systemic pesticides on biodiversity and ecosystems. Environmental Science and Pollution Research. 2015;22(1):1-54.

[74] Nakanishi K, Yokomizo H, Hayashi TI. Were the sharp declines of dragonfly populations in the 1990s in Japan caused by fipronil and imidacloprid? An analysis of Hill's causality for the case of Sympetrum frequens. Environmental Science and Pollution Research. 2018 Dec;25 (35):35352-35364.

[75] Strong L. Avermectins: A review of their impact on insects of cattle dung. Bulletin of entomological research. 1992 Jun;82(2):265-274

[76] McKenna DD, McKenna KM, Malcom SB, Bebenbaum MR. Mortality of Lepidoptera along roadways in Central Illinois. Journal-lepidopterists society. 2001 Dec 1;55(2):63-68.

[77] Seibert HC, Conover JH. Mortality of vertebrates and invertebrates on an Athens County, Ohio, highway.

[78] Stork NE. How many species of insects and other terrestrial arthropods are there on earth?. Annual review of entomology. 2018 Jan 7;63:31-45.

[79] Møller AP. Parallel declines in abundance of insects and insectivorous birds in Denmark over 22 years. Ecology and evolution. 2019 Jun;9(11):6581-6587.

[80] Bell JR, Blumgart D, Shortall CR. Are insects declining and at what rate? An analysis of standardised, systematic catches of aphid and moth abundances across Great Britain. Insect conservation and diversity. 2020 Mar;13(2):115-126.

[81] McInnes, D. 2021. Florida Museum of Natural History. "unsure how to help reverse insect declines? Scientists suggest simple ways." ScienceDaily. www.sciencedaily.com/releases/2021/01/210112110144.htm

[82] Didham RK, Basset Y, Collins CM, Leather SR, Littlewood NA, Menz MH, Müller J, Packer L, Saunders ME, Schönrogge K, Stewart AJ. Interpreting insect declines: Seven challenges and a way forward. Insect Conservation and Diversity. 2020 Mar;13(2):103-114

[83] Sandom C, Faurby S, Sandel B, Svenning JC. Global late quaternary megafauna extinctions linked to humans, not climate change. Proceedings of the Royal Society B: Biological Sciences. 2014 Jul 22;281 (1787):20133254.

[84] Vargas-Larreta B, López-Martínez JO, González EJ, Corral-Rivas JJ, Hernández FJ. Assessing above-ground biomass-functional diversity relationships in temperate forests in northern Mexico. Forest Ecosystems. 2021 Dec;8(1):1-4.

[85] Boyes DH, Fox R, Shortall CR, Whittaker RJ. Bucking the trend: the diversity of Anthropocene 'winners' among British moths. Frontiers of Biogeography. 2019;11(3).

[86] Dirzo R, Young HS, Galetti M, Ceballos G, Isaac NJ, Collen B. Defaunation in the Anthropocene. science. 2014 Jul 25;345(6195): 401-406.

[87] Fox R, Harrower CA, Bell JR, Shortall CR, Middlebrook I, Wilson RJ. Insect population trends and the IUCN red list process. Journal of Insect Conservation. 2019 Apr;23(2):269-278.

[88] Stockton DG, Wallingford AK, Loeb GM. Phenotypic plasticity promotes overwintering survival in a globally invasive crop pest, Drosophila suzukii. Insects. 2018 Sep;9(3):10

[89] Dornelas M, Daskalova GN. Nuanced changes in insect abundance. Science. 2020 Apr 24;368(6489):368

[90] Crossley MS, Smith OM, Berry LL, Phillips-Cosio R, Glassberg J, Holman KM, Holmquest JG, Meier AR, Varriano SA, McClung MR, Moran MD. Recent climate change is creating hotspots of butterfly increase and decline across North America. Global Change Biology. 2021 Mar 22.

[91] Raven PH, Wagner DL. Agricultural intensification and climate change are rapidly decreasing insect biodiversity. Proceedings of the National Academy of Sciences. 2021;118(2)

[92] Patrick CJ, McCluney KE, Ruhi A, Gregory A, Sabo J, Thorp JH. Multi-scale biodiversity drives temporal variability in macrosystems. Frontiers in Ecology and the Environment. 2021 Feb;19(1):47-56.

[93] Soroye P, Newbold T, Kerr J. Climate change contributes to widespread declines among bumble bees across continents. Science. 2020 Feb 7;367(6478):685-688.

[94] Senapathi D, Fründ J, Albrecht M, Garratt MP, Kleijn D, Pickles BJ, Potts SG, An J, Andersson GK, Bänsch S, Basu P. Wild insect diversity increases inter-annual stability in global crop pollinator communities. Proceedings of the Royal Society B. 2021 Mar 31;288 (1947):20210212.

[95] Outhwaite CL, Gregory RD, Chandler RE, Collen B, Isaac NJ. Complex long-term biodiversity change

among invertebrates, bryophytes and lichens. Nature ecology & evolution. 2020 Mar;4(3):384-392.

[96] Wang S, Brose U. Biodiversity and ecosystem functioning in food webs: The vertical diversity hypothesis. Ecology Letters. 2018 Jan;21(1):9-20.

[97] Benton TG, Bieg C, Harwatt H, Pudasaini R, Wellesley L. Food System Impacts on Biodiversity Loss. Three levers for food system transformation in support of nature. Chatham House, London. 2021 Feb.

[98] Ritchie H. 2021. Our world in data. Extinctions. Global change data Lab. England and Wales Charity trust Reg. No 1186433. https://ourworldindata.org/

[99] Ceballos G, Ehrlich PR, Raven PH. Vertebrates on the brink as indicators of biological annihilation and the sixth mass extinction. Proceedings of the National Academy of Sciences. 2020 Jun 16;117(24):13596-13602.

[100] Schachat SR, Labandeira CC. Are insects heading toward their first mass extinction? Distinguishing turnover from crises in their fossil record. Annals of the Entomological Society of America. 2020 Dec 31.

[101] de Lima RA, Oliveira AA, Pitta GR, de Gasper AL, Vibrans AC, Chave J, Ter Steege H, Prado PI. The erosion of biodiversity and biomass in the Atlantic Forest biodiversity hotspot. Nature communications. 2020 Dec 11;11(1):1-6.

[102] McCosker J, Smith DG, Tighe K, Torres AG, Leander NJ. *Muraenesox cinereus*. The IUCN Red List of Threatened Species 2021: e. T199344A2 585390.

[103] Schowalter TD. Insect Ecology: An Ecosystem Approach. Academic press; 2016 Jul 29

[104] Bianchi FJ, Booij CJ, Tscharntke T. Sustainable pest regulation in agricultural landscapes: A review on landscape composition, biodiversity and natural pest control. Proceedings of the Royal Society B: Biological Sciences. 2006 Jul 22;273(1595):1715-1727.

[105] Biesmeijer JC, Roberts SP, Reemer M, Ohlemüller R, Edwards M, Peeters T, Schaffers AP, Potts SG, Kleukers RJ, Thomas CD, Settele J. Parallel declines in pollinators and insect-pollinated plants in Britain and the Netherlands. Science. 2006 Jul 21;313(5785):351-354

[106] Klein AM, Vaissiere BE, Cane JH, Steffan-Dewenter I, Cunningham SA, Kremen C, Tscharntke T. Importance of pollinators in changing landscapes for world crops. Proceedings of the royal society B: biological sciences. 2007 Feb 7;274(1608):303-313

[107] Habel JC, Schmitt T. Vanishing of the common species: Empty habitats and the role of genetic diversity. Biological Conservation. 2018 Feb 1;218:211-216.

[108] Hochkirch A, Schmitt T, Beninde J, Hiery M, Kinitz T, Kirschey J, Matenaar D, Rohde K, Stoefen A, Wagner N, Zink A. Europe needs a new vision for a Natura 2020 network. Conservation Letters. 2013 Nov;6(6):462-467

[109] Habel JC, Segerer A, Ulrich W, Torchyk O, Weisser WW, Schmitt T. Butterfly community shifts over two centuries. Conservation Biology. 2016 Aug;30(4):754-762.

[110] Hewitt JE, Thrush SF, Ellingsen KE. The role of time and species identities in spatial patterns of species richness and conservation. Conservation Biology. 2016 Oct;30(5):1080-1088.

[111] Garibaldi LA, Pérez-Méndez N, Garratt MP, Gemmill-Herren B, Miguez FE, Dicks LV. Policies for ecological intensification of crop

production. Trends in ecology & evolution. 2019 Apr 1;34(4):282-286.

[112] Ranjha MH, Irmler U. Movement of carabids from grassy strips to crop land in organic agriculture. Journal of insect conservation. 2014 Jun;18(3):457-467.

[113] Toivonen M, Herzon I, Kuussaari M. Community composition of butterflies and bumblebees in fallows: Niche breadth and dispersal capacity modify responses to fallow type and landscape. Journal of Insect Conservation. 2016 Feb;20(1):23-34.

[114] Woodcock BA, Bullock JM, Shore RF, Heard MS, Pereira MG, Redhead J, Ridding L, Dean H, Sleep D, Henrys P, Peyton J. Country-specific effects of neonicotinoid pesticides on honey bees and wild bees. Science. 2017 Jun 30;356(6345):1393-1395.

[115] Rundlöf M, Nilsson H, Smith HG. Interacting effects of farming practice and landscape context on bumble bees. Biological conservation. 2008 Feb 1;141(2):417-426.

[116] Rundlöf M, Nilsson H, Smith HG. Interacting effects of farming practice and landscape context on bumble bees. Biological conservation. 2008 Feb 1;141(2):417-426.

[117] Seufert V, Ramankutty N, Foley JA. Comparing the yields of organic and conventional agriculture. Nature. 2012 May;485(7397):229-232.

[118] Bakacs ME, Yergeau SE, Obropta CC. Assessment of car wash runoff treatment using bioretention mesocosms. Journal of Environmental Engineering. 2013 Aug 1;139(8): 1132-1136.

[119] Koen EL, Minnaar C, Roever CL, Boyles JG. Emerging threat of the 21st century lightscape to global biodiversity. Global change biology. 2018 Jun;24(6):2315-2324

[120] Frank KD, Rich C, Longcore T. Effects of artificial night lighting on moths. Ecological consequences of artificial night lighting. 2006:305-344.

[121] Coulthard E, Norrey J, Shortall C, Harris WE. Ecological traits predict population changes in moths. Biological Conservation. 2019 May 1;233:213-219.

[122] Elgert C, Hopkins J, Kaitala A, Candolin U. Reproduction under light pollution: Maladaptive response to spatial variation in artificial light in a glow-worm. Proceedings of the Royal Society B. 2020 Jul 29;287(1931):20200806.

[123] Hussain, B., Ahmad, B. and Bilal, S., 2015. Monitoring and mass trapping of the codling moth, Cydia pomonella, by the use of pheromone baited traps in Kargil, Ladakh, India. International Journal of Fruit Science, *15*(1), pp.1-9.

Chapter 2

# Potential Reasons for Insect Decline

*Gagan Preet Kour Bali and Amritpal Singh Kaleka*


## Abstract

Insects are the key component of world's ecosystem and act as vital force to maintain life's framework. But in present scenario, Insects are under multi-continental crisis apparent as reduction in abundance, diversity and biomass. The impact of decline is severe in areas which are highly impacted by human activities such as industrialized and agricultural landscapes. Habitat loss and degradation; intensive use of pesticides; pollution; introduction of invasive species and climate change are the most influential factors for their alarming decline and each factor is multifaceted. The accelerated decline in insect population can cause unpredictable negative consequences for the biosphere and is a matter of global concern that requires immediate and effective international collaborations. An urgent need is to identify the species at greatest threat; factors threatening their survival and finally the consequences of their loss. In order to maintain the integrity of managed and natural ecosystems, the protection of Insect diversity is critically important.


**Keywords:** Climate change, ecosystem, habitat loss, industrialization, invasive species

## 1. Introduction

Insects are cosmopolitan in distribution i.e., found in every possible environment. They have adapted to a broad range of habitats, successfully finding their own niche as these organisms feed on any substance that has nutritional value. They constitute crucial component of environment and act as key components in the functioning of world's ecosystems. Their accelerated decline in numbers and extinction due to anthropogenic activities cause unpredictable negative consequences for the biosphere. Insects dominate the entire globe and have silently witnessed the rise of vertebrates, the fall of non-avian dinosaurs, the proliferation of mammals and the rapid evolution, civilization and industrialization of humans. Insects are so diverse that their numbers are impressive even in most parochial places. The degree of success achieved by a group of organisms can be measured either as total number of individuals within the group or more commonly as the number of different species of organisms that comprise the group. On both accounts, the insects must be considered highly successful and established group. Their success depends on two interacting factors:

- The potential of the group for adapting to new environmental conditions.
- The degree to which the environmental conditions change.

Insects create the biological foundation for all terrestrial ecosystems by cycling nutrients, pollinating plants, dispersing seeds, maintaining soil structure and fertility, controlling populations of other organisms and by providing major food source for other taxa. Once the benefits of insects-provided services are realized there may be start up call for increased funding to conserve insects through legislation. Insects are certainly under represented and underfunded through legislation and increased funds could save insect species from extinction. Because insects constitute the world's most abundant and speciose animal group and provide critical services within ecosystems, cannot be ignored and prompt a decisive action to avert a catastrophic collapse of nature's ecosystem [1].

## 2. Insect decline: a major concern

> *"Bugs nevrer bug my head. They are amazing. It is the activities of humans which actually bug me all the time." (Munia Khan)*

Declines have been severe in areas highly impacted by human activities such as industrialized and agricultural landscapes but ongoing insect declines are not restricted to the farms or footprints of suburban sprawl. The reduction in total biomass in multi-decadal studies is similarly being reported from different parts of the globe. About 33% reduction in the abundance of butterflies was observed over a span of 21 yrs. in extensive monitoring in Ohio, USA [2]; 176 moth species decreased by 20% from 1975 to 2014 in Rothamsted insect survey in Scotland [3]; total flying insect biomass decreased by more than 70% across 63 study locations over 27 yrs. in Germany [4].

Declines have not only been observed among species with narrow habitat requirements but also among those which are broadly distributed and abundant. Anthropogenic pressure is shifting multiple insect communities towards species-poor assemblages and the affected insect groups not only include specialists that occupy particular ecological niches, but also many common and generalist species [5]. The current biodiversity losses and shifts in community composition could cause the extinction [6]. Species losses are expected to lead to a steady decay of insect mediated ecosystem services, which are likely to be provided by fewer and less specialized species [7, 8]. Five massive extinctions of life on earth have already occurred in the distant geological past, it is considered that the present day biodiversity extinction will be the largest in the history of life [9]. The extinction rate due to anthropogenic causes such as habitat destruction, overharvesting, pathogens, pesticides, addition of pollutants, urbanization, inclusion of invasive species and emission of greenhouse gases will be probably thousands times larger than the background rate [10]. Insects are not immune to this unprecedented wave of extinctions due to the reasons indicated above; they have also been neglected in relation to other more charismatic species [11, 12].

Despite the ubiquity of insects and their extensive connections to plants and other animals, declines in insect diversity and abundance are apparent in studies including faunal and biomass assessments and status reviews of key indicator groups [13].

## 3. Loss of abundant species

Conservation efforts have mainly focused their attention on protecting and conserving rare, charismatic and endangered species but the "Insect apocalypse" presents a different challenge. The sweeping declines of formerly abundant insects in addition to the loss of rare taxa have raised concern about ecosystem functioning [14].

The insect declines potentially have created global ecological and economic consequences. Not all insects are declining and many lineages have not changed rather increased in abundance [15, 16]. For instance: In Great Britain many species of moths have expanded in their range or population size [17]; numerous temperate insects, presumably limited by winter temperatures, have increased their abundance and range, in response to warmer global temperatures [14]; anthropophilic and human-assisted taxa, which include many pollinators, such as the western honey bee (*Apis mellifera* Linnaeus) in North America, thrive well due to their associations with humans; abundance of freshwater insects attributed to clean water legislation, in both Europe and North America [18]. In some places, native herbivores have flourished by utilizing nonnative plants as adult nectar sources or larval food plants [19], and there are even instances where introduced plants have rescued imperiled species.

## 4. Systematic drivers of insect decline

Despite great diversity of ecologies and life histories represented by insects in different regions and habitats, patterns are emerging that point to the primary drivers of insect declines. Most influential factors are habitat degradation and loss, excessive pesticide usage and climate change [20], although other factors include diseases, invasive species and pollution. Though Drizo *et al.* considered the insect decline due to multiplicity of factors, but they termed habitat destruction, deforestation, fragmentation, urbanization and agricultural conversion as the leading factors responsible in insect decline [21]. They captured the essence of the problem that the Insects are suffering from "death by thousand cuts" (**Figure 1**).

The potential reasons for Insect decline are as under:

- Habitat Loss
- Intensive use of pesticides and herbicides
- Pollution
- Introduction of invasive species
- Climate change

### 4.1 Habitat loss

Undoubtedly, the most serious cause of insect decline is their habitat loss. It takes place as natural habitats are being converted to human utilization areas such as croplands i.e., agriculture areas, urban areas and for infrastructure development (e.g. roads, industries, dams, power stations etc). The habitat loss eventually results in species extinction and loss of biodiversity. It is not an exclusively man-made phenomenon, habitat loss also occurs as a result of natural events such as floods, volcanic eruptions, earthquakes and climatic fluctuations.

Habitat loss can be broadly categorized into three major types:

a. Habitat destruction

b. Habitat degradation

c. Habitat fragmentation

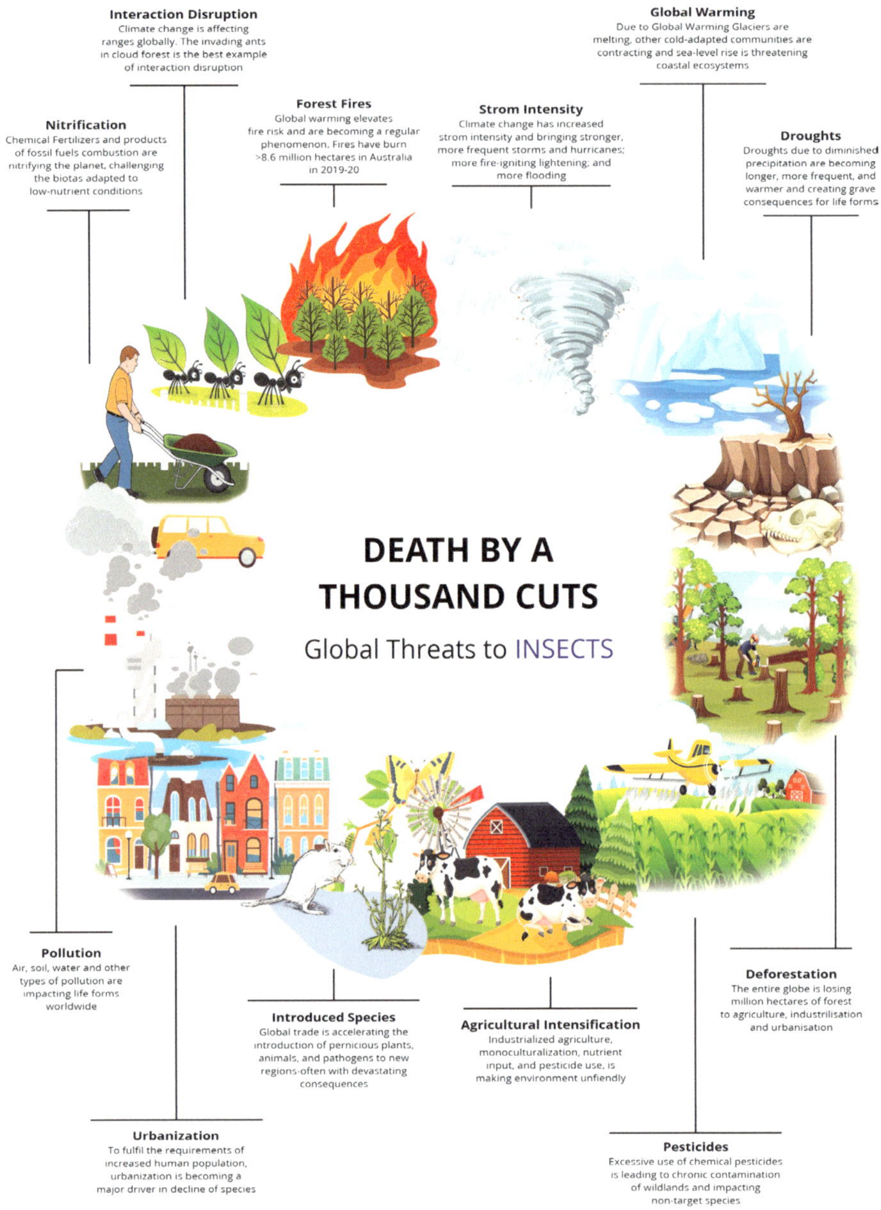


**Figure 1.**
*Death by a thousand cuts: Major drivers responsible for insect decline [21].*

**(a) Habitat destruction**: It involves the processes by which natural habitat is destroyed or damaged as it remains no longer capable of supporting the species and ecological communities occurring naturally in that area. It results in extinction of species and ultimately loss in biodiversity at large [22]. Human activities such as clearing of land for various purposes like agriculture, mining, logging, hydroelectric projects, urbanization etc. directly destruct the habitats. It leads to species extinction but on the contrary opens up new opportunity habitats for evolution of new species, thus demonstrating the resilience of life on earth. Unfortunately, most species and communities are not able to cope with these changes as natural habitats are being destroyed at a much higher rate and spatial scale [23].

**(b) Habitat degradation**: The factors such as pollution, introduction of invasive species and over utilization of natural resources are causing decline in the biological conditions. This decline in biological conditions is further degrading the natural habitats. The habitat degradation reduces the quality of the environment, making it difficult for native plants and animals to thrive [24]. Habitat degradation is fueled by fast-growing human population. As the population increases, humans use more land for agriculture and for development of cities and towns which spread out over ever-widening areas. The effects of habitat degradation not only affect native species and communities but human populations as well. The degraded lands are frequently lost to erosion, desertification and nutrient depletion.

**(c) Habitat or Landscape fragmentation**: It involves the breakage up of a habitat or vegetation type into smaller, disconnected patches. The major consequence of land use involving agriculture activities, construction of roads, housing projects etc. are resulting in the fragmentation of existing habitats. This process involves three major steps i.e. Landscape dissection; Landscape perforation and Landscape attrition. Fragmentation reduces animal ranges and restricts their movement. It poses higher risks of extinction and decreases genetic diversity among them. It results in reduction of habitat area which automatically leads to:

- Increased Isolation i.e., habitat patches are no longer connected to each other which results in loss of biodiversity. Some species temporarily disappear and the net result is lower number of species.

- Smaller habitat patches

- Negative and positive edge effects

It is generally accepted among conservation biologists that the ongoing fragmentation and reduction in area of natural habitats is causing species extinctions at local, regional and global levels. The remaining areas of more-or-less natural habitats are increasingly becoming mere pockets with in a sea of habitats or in other words as habitat islands.

Several butterflies breed in the canopy (feeding on deciduous trees) but other in open spaces like clearings, glades, shrubs, hedges etc. The open spaces are also declining due to lack of management or re-plantation of coniferous trees which cast shade. Certain species namely *Boloria euphrosyne* (Linnaeus) and *Boloria selene* (Denis & Schiffermüller) have drastically declined in number. The populations breeding in fragmented habitats are likely to become extinct, through normal stochastic processes or by inbreeding depression. Hanski discussed about the butterflies which have been reported to occur frequently as meta-populations spanning in small patches of habitat [25]. The species depending on particular resources such as specific food plants for larval development or specific microhabitats are mainly affected by deteriorating habitat quality. However, sedentary species suffer more severely under the driver of increasing habitat isolation [26]. More extinctions are sure to occur if the loss of natural habitat around the globe does not slow or pace down [27].

### 4.2 Intensive use of pesticides and herbicides

The excessive use of fertilizers and synthetic pesticides in agricultural practices constitutes the second major driver of insect decline. The pesticide pollution is reported in 13% of cases, followed by fertilizer inputs (10%) and to a lesser extent by urban and industrial pollutants (3%).The systematic and widespread use of pesticides for controlling crop pests (insecticides), competing weeds (herbicides) and fungal infections (fungicides) forms the basis of modern intensive agriculture [28]. Insecticides are most toxic to insect life and other arthropods, followed by fungicides and then herbicides in terms of toxicity [29]. The herbicide application to croplands have more negative impact on both terrestrial and aquatic plants and insect diversity than any other agricultural practice [30]. The herbicides reduce the vegetation diversity within the crops and runoff, thus impacting indirectly on the arthropod species which are dependent on wild plants. This reduction results in either complete disappearance or significant decline in their numbers [31].

Insecticides such as pyrethroids, neonicotinoids and fipronils have devastating impact on aquatic insects and crustaceans due to their high acute and chemical toxicity [32], thus reducing their abundance significantly in water bodies. Neonicotinoids are known to kill monarch butterflies in the laboratory conditions and their lethal quantities are found in host plants of these butterflies in the fields [33]. These pesticides are persistent and leach into the soil and water courses as well as into field margins where certain butterfly species breed or forage. Neonicotinoids also play a major role in the decline of bees [34]. The exposures to even low doses of pesticides have complex and unpredictable sub-lethal impact on insect behavior. Bees when exposed to these pesticides get confused and are unable to find their way back to the hive. Even minute amount of neonicotinoids i.e., 1 part per billion in food impairs their immune system thus making them susceptible to diseases like deformed wing virus etc. Nakanishi *et al.* discussed about neonicotinoids as one of the major drivers of dragonfly decline in Japan [35]. The pesticides caused the decline in moth numbers in the rural areas of U.K [36] and pollinators in Italy [37]. Lundgren *et al*. concluded that the broad spectrum insecticides reduce the abundance and diversity of beneficial, ground dwelling and foliage foraging insects [30]. The systemic insecticides reduce population of ladybirds and butterflies in gardens and nurseries [38] and inflict multiple lethal and sub-lethal effects on bees and other arthropods. The residues of pesticides namely fipronil in sediments inhibit the emergence of dragonflies [39] and the development of chironomids and other insect larvae thereby having cascading effects on fish survival [40]. Hallamann *et al.* [41] demonstrated that 80% of the flying insect biomass losses in Germany were not caused by increase in agricultural land, deforestation or climate change but occurred due to intensive use of pesticides.

### 4.3 Pollution

Pollution covers a wide variety of substances that adversely influence insect fauna and their habitats.

The aerial nitrogen deposition is a harmful pollutant affecting butterflies. The ammonia produced by intensive livestock rearing and the emission of nitrogen oxides from vehicles are the chief sources of nitrogen pollution. It either changes the microclimate [42] or the nature of the vegetation where the butterflies breed. Nitrogen deposition encourages vegetation growth and reduces the amount of bare ground where *Lasiommata megera* (Linnaeus) commonly known as the wall or wall brown butterflies breed in Netherlands [43]. Nitrogen accumulation also leads to replacement of flowering herbs by grasses [44]. It reduces the availability for flower visiting insects and specifically larval food plants for many phytophagous species. Pollinators are adversely affected by the decline in flowering herbs [45]. Nitrogen deposition also results in cooling off warm microclimates resulting in the decline of butterfly species which overwinter as eggs or caterpillars rather than as adults or chrysalids [42]. This cooling off reduces the growth rate and chances of survival of such insects.

Light pollution also acts as driver of insect decline in suburban and urban locations. It has a significant impact on nocturnal insects such as moths [46]. The adoption of artificial lighting at night (ALAN) is a growing threat to biodiversity in general and particularly to nocturnally active insects. The ALAN have impact on vital behaviors of nocturnally active insects, including feeding, migration and dispersal, predator avoidance and reproduction, potentially with cascading effects on the diversity of insects and the ecosystem services that they provide. Brehm *et al.* [47] tested free flying individuals of 95 moth species with a choice of specific light wavelengths under controlled conditions. They observed that attractiveness increased with both light intensity and the shorter wavelengths. The insects that produce their own light i.e., bioluminescent insects of family Lampyridae referable to order Coleoptera variously known as fireflies, glow worms or lightening bugs for signaling are severely affected by ALAN. It distracts or disorientates either the courtship partners or simply reduces the efficacy of communication by flooding the background with illumination thereby decreasing the signal to noise ratio. Deichmann *et al.* [48] suggested that switching to lights of longer wavelength in order to minimize the adverse effects on majority of insects in turn attract bioluminescent insects thereby posing threat to their survival. Langevelde *et al.* [49] studied the effect of artificial light on feeding behavior of moths. They observed that the moths subjected to artificial night lighting spend less time in feeding than in darkness, with shortest time under light conditions rich in short wavelength. Boyes *et al.* [50] suggested that the diurnal adult stages of Lepidoptera are indirectly affected by the impacts of ALAN on their nocturnal feeding caterpillars. Therefore, restoration and maintenance of darkness in illuminated areas is essential for reversing decline in moth populations. Kalinkat *et al.* [51] highlighted the scarcity of evidence that ALAN has made any significant contribution to decline in insects.

The biggest future challenge is the assessment and documentation of impacts of ALAN on individual insects and their detectable effects over long time scale on the dynamics of populations, communities and ecosystems.

### 4.4 Introduction of invasive species

Invasive species are among the largest threats to biodiversity in the world. According to International Union for Conservation of Nature (IUCN) an invasive species is an alien species which becomes established in natural or semi-natural ecosystems or habitats. It is an agent of change and threatens native biological diversity. Invasive species possess specific traits or combination of traits that allow them to outcome native species. These species tend to have the following traits [52]:

- Fast growth
- Rapid reproduction
- High dispersal ability
- Phenotypic plasticity i.e., the ability to alter growth form to suit the current conditions
- Tolerance of a wide range of environmental conditions- Ecological competence
- Ability to live off on a wide range of food types

- Association with humans
- Prior successful invasions

According to Rejmanek & Richerdson [53] invasive species tends to be hardy with long life span, voracious feeding habits, aggressively pervasive, very resilient, rapid growth, generalized diet, ability to move long distances and most significant is its profilic breeding. Although, these species pose a substantial threat to biodiversity but may also increase evolutionary diversification due to expansion of geographic range, increase in number of generations, breakdown of host plant resistance etc.

In ecosystem, predators, herbivores and other wildlife evolve alongside each other, regulating each other's populations. But a non-native species can disrupt that balance and wipe out organisms resulting in large populations of the invasive species. Invasive species have profoundly reduced biodiversity in some ecosystems. Human mediated redistribution i.e., both deliberate and accidental distribution of insect species has led to decline in many native species through competition or with displacement by invasive species. The accidental introduction of an Argentine ant species into the unique vegetation community of the Cape Province of South Africa is a notable example. It has led to the decline of indigenous ant species adapted to disperse the seeds of many plants. The reliably identified causes showed that the invasive species contributed directly to the demise of 91 (54%) of 170 extinct species. Particularly, the rates of extinction occurring on islands have been greatly elevated by the introduction of novel predators. Several ecological and life history attributes of island species, such as their naturally constrained geographic range, small population size and particular traits make island biota vulnerable to predation from invading species [54].

A recent study pointed out that an increase in the spread of non-native plant and animal species around the world could lead to dramatic biodiversity loss, causing permanent damage to ecosystems. Prof. Helen Roy of UK Centre for Ecology & Hydrology said: "With invasions, it's not that were trying to return to some kind of pristine environment or some kind of norm, but it's around the functioning of those ecosystems. And that's we need to have a much better understanding of."

### 4.5 Climate change

Over the course of time, there have been repeated cycles of climate changes and these changes have driven massive alterations in the distribution of species across the Globe. Individual species have experienced alternating episodes of expansion, contraction and fragmentation of ranges. Now, it is the humans who are driving the range shifts and extinctions and doing so in an accelerating fashion on a global scale. Human mediated climate change represents a potentially disastrous sleeping giant in terms of future biodiversity losses. Climate warming can affect species in five principal ways:

- Alterations of species densities including altered community composition and structure
- Range shifts either pole-ward or upward in elevation
- Behavioral changes such as phenology i.e., changes in seasonal timing of life cycle events of migration, breeding and flowering

- Changes in morphology such as body size
- Reduction in genetic diversity that leads to inbreeding depression

Climatic changes alter almost every aspect of plant and animal diversity and particularly insect diversity. The significant alterations may be summarized into following categories or responses [55].

1. ORGANISMAL RESPONSES
    - Genetic responses
    - Behavioral responses
    - Morphological responses
2. POPULATION-LEVEL RESPONSES
    - Population and range expansion
    - Population and range retraction
3. PHENOLOGICAL RESPONSES
    - Voltinism
    - Early emergence and asynchronies
4. COMMUNITY RESPONSES
    - Trophic mismatches

Climate changes have both positive and negative effects on insects especially butterflies. The rapid growth of natural vegetation and the problems associated with intermediate successional vegetation due to climatic change affects the microclimate or the nutritional quality of food plants [56]. The dense vegetation favors drought conditions leading to hot, frequent fires having detrimental effects on insect taxa which are not adapted to high temperature conditions [57]. Multi-voltine species are declining less rapidly than uni-voltine ones in UK [58] whereas the opposite is the case in Mediterranean region of Spain [59]. Phenological plasticity is observed in many species as an indicator of resilience to climate change. The species with complex life cycles i.e., holometabolus insects particularly butterflies suffer from developmental traps if environment cues to enter diapause are disrupted e.g. *Lasiommata megeara* Linnaeus (Wall brown butterfly) in Belgium [60].

Climate change is predicted to increase the frequency of extreme weather events like droughts and floods. Droughts lead to rapid decline in insect populations particularly butterfly abundance. Prolonged rainfall and storms reduce breeding success resulting in reduction of overall population size. Schowalter *et al.* [61] concluded that insect responses to temperature within Puerlo Rico's Luquillo experimentation forest, a hurricane mediated ecosystem are driven principally by storms and post-storm effects rather than by global climate warming. The geographical ranges of some insects have started to shift in response to climate. The European and North American bumblebees tend to disappear from the southern edges of their range [62] and occupied higher elevations in mountainous region [63]. Bark beetles in North America have become more abundant due to warmer winters resulting in defoliation of coniferous forests. The increased frequency, intensity and duration of extreme weather condition events disrupted food webs, producing seasonal mismatches between specialized insect pollinators and the plants associated with such insects. Climate change especially warming temperature makes conflict between morphological traits such as dispersal. The selection for reduced body size in warmer environment leads to associated loss of dispersal capacity. Wu *et al*. [64] studied the reduction of wing size in Bornean geometrid moths at high altitude and linked it to uphill shifts of smaller species after four decades of warming. This unequal redistribution of different sized species significantly affects community size composition.

Genetic variation allows a species to develop tolerance in different environments as selection acts on dispersal capacity [65]. While studying the thermoregulation and behavior of lowland species of a cold-dwelling butterfly genus *Erebia Dalman,* Kleckova and Klecka concluded that the selective pressure on butterfly populations altered due to anthropogenic mediated climatic changes and led to allele frequency shifts associated with dispersal [66]. Genetic polymorphism in *Pgi* gene encoding phosphoglucose isomerase influences key history traits in adult insects including dispersal, flight metabolism, longevity and fecundity [67]. *Pgi* heterozygote butterflies have increased fitness in cooler climates as they are capable of flying at lower ambient temperature than their homozygote counterparts [68]. In *Araschina levana* (Linnaeus) (map butterfly) individuals have higher levels of dispersive *Pgi* alleles at new colonized sites even though no morphological changes which improve flight performance such as increased wing or thorax size are seen [69]. *Pgi* alleles are also associated with heat resistance in Lepidoptera. In *Colias* butterflies, the genotypes that are most heat stable have low fecundity, so selection for heat tolerance greatly reduces population sizes [70]. Another molecular marker important in response to climate change is the heat shock protein (Hsp70). It plays a critical role in helping insects to survive in extreme temperature by increasing tolerance [71]. Both *Pgi* & Hsp70 offer a robust comparison of key genes and phenotypes directly impacted by changing climate.

All the forces or factors may act independently or synergistically and thus identification of a single cause of a particular species extinction event is difficult. For instance, habitat loss may cause some extinction directly by removing all individuals, but it can also be indirectly responsible for extinction by facilitating the establishment of an invasive species or disease agent, improving access to human hunters, or altering biophysical conditions. As a result, any process that causes a population to dwindle may ultimately predispose that population to extinction. When climate change and habitat loss act synergistically with each other, leads to a deadly anthropogenic cocktail which is more deadly when there is an increase in intensity at the same time [72].

The cognitive abilities of honeybees have found to be impaired by low electromagnetic fields such as those created around high-voltage cables. Shepherd *et al.* concluded that this has contributed to bee colony losses and more broadly could impact on insect navigation and dispersal [73]. It seems that there are other human activities which affect insect health in many ways and are yet to be recognized and assessed by scientists for their impact on the insects and their environment.

## 5. Future steps in conservation strategies: a call to action

Being the major constituents of biodiversity, insects have high ecological and economic importance. These creatures play key roles in species interactions and constitute a major component in all food webs to provide resources for organisms at higher trophic levels [61]. Insect decline is not the prime matter of concern but highly important for well-being of humanity. The necessary conservation activities needed to counteract main drivers of insect decline are also equally important. The following steps are required for developing conservation strategies of insect fauna:

1. Conservation of high quality habitats: The first and foremost priority is to protect high quality insect habitats. For this, high quality habitats in the agricultural matrix have to be reestablished with extended size as small and isolated nature reserves are not able to guarantee long lasting preservation of insect species [74].

2. Increasing landscape permeability: Healthy population network with high functional connectivity is the main reason behind long term persistence of many species [75]. The government authorities should provide economic incentives for ecosystem conservation and such incentives help to stop landscape fragmentation, in creation of additional high quality habitats and improvement in quality of existing habitats such as reversion of monotonous, high productivity grasslands into diverse flower rich meadows. The ecological intensification of agriculture e.g. field margin extension and roadside ecological landscaping increase habitat connectivity in an area of flowering plants and improves the landscapes to insect friendly conditions; grassy strips encourage ground beetles to move into adjacent fields [76]; small temporal fallows of arable fields improve condition for bumble bees and butterflies. It acts as long term insurance policy for future delivery of irreplaceable and essential insect services.

3. Safeguarding habitat quality: The detrimental effects of pesticides and herbicides have to be reduced both in agricultural and urban arenas particularly neonicotinoids which reduce the capacity of bee species to establish new populations [77]. The chemicals known to strongly harm insect diversity even in sub lethal doses should be banned. Organic farming practices along with in field plant diversification greatly benefits insect fauna especially pollinators.

The insect decline and conservation have to be understood as a societal and economic challenge along with scientific concern [78]. It requires six basic requisites, all on economic viable platform i.e.,

- Philosophy (establishing the ethical foundation)

- Research (the finding out)

- Policy (framework for action)

- Psychology (understanding how to engage humans in insect conservation action)

- Practice (implementation of action)

- Validation (establishing how well we are doing at conserving insects)

There must be a coordinated effort among scientists, NGO's, policy makers, funding agencies, science communicators and citizens around the globe to find solutions to curb decline in insect diversity and abundance.

## 6. Conclusion

Insects act as key components for the functioning of the world's ecosystems. Insects create the biological foundation in all ecosystems by cycling nutrients, pollinating plants, dispersing seeds, maintaining soil structure and fertility, controlling populations of other organisms and by providing major food source for other taxa. Their accelerated decline in numbers and extinction due to anthropogenic activities cause unpredictable negative consequences for the biosphere. Declines have not only been observed among species with narrow habitat requirements but also among those that are broadly distributed and abundant. Anthropogenic pressure is shifting multiple insect communities towards species-poor assemblages dominated by experts [79]. The falling number of insect populations is likely due to a multiplicity of factors, habitat destruction, deforestation, fragmentation, urbanization and agricultural conversion being among the leading factors. All the forces or factors may act independently or synergistically and thus identification of a single cause of a particular species extinction event is difficult.

Insect decline is not the prime matter of concern but the necessary conservation activities required to neutralize main drivers responsible for insect decline are mainly important for well-being of humanity. Not only the Government and legislative actions are required, even the action of individuals can create immediate impact. It is justified by a simple fact that conserving even a backyard or apartment balcony can be an important stopover for the smallest insect groups upon which we all depend.

**Author details**

Gagan Preet Kour Bali* and Amritpal Singh Kaleka
Department of Zoology and Environmental Sciences, Punjabi University, Patiala, Punjab, India

*Address all correspondence to: gaganviren@gmail.com

## References

[1] May RM. Ecological science and tomorrow's world. Philosophical Transactions of the Royal Society London B: Biological Sciences. 2010; **365**:41-47.

[2] Wepprich T, Adrion JR, Ries L, Wiedmann J, Haddad NM. Butterfly abundance declines over 20 years of systematic monitoring in Ohio, USA. *BioRxiv*. 2019;613786.

[3] Dennis EB, Brereton T, Morgan BJT, Fox R, Shortall CR, Prescott T, Foster S. Trends and indicators for quantifying moth abundance and occupancy in Scotland. Journal of Insect Conservation. 2019; **23**(2): 369-380.

[4] Hallmann CA, Sorg M, Jongejans E, Siepel H, Hofland N, Schwan H, Stenmans W, Müller A, Sumser H, Hörren T, Goulson D, de Kroon H. More than 75 percent decline over 27 years in total flying insect biomass in protected areas. PLos One. 2017; **12**: e0185809.

[5] White PJT, Kerr JT. Human impacts on environment diversity relationships: evidence for biotic homogenization from butterfly species richness patterns. Global Ecology and Biogeography. 2007; **16**: 290-299.

[6] Chapin-III FS, Zavaleta ES, Eviner VT, Naylor RL, Vitousek PM, Reynolds HL, Hooper DU, Lavorel S, Sala OE, Hobbie SE, Mack MC, Diaz S. Consequences of changing biodiversity. Nature. 2000; **405**: 234-242.

[7] Bartomeus I, Potts SG, Steffan DI, Vaissière BE, Woyciechowski M, Krewenka KM, Tscheulin T, Roberts SPM, Szentgyörgyi H, Westphal C, Bommarco R. Contribution of insect pollinators to crop yield and quality varies with agricultural intensification. Peer J. 2014; **2**: e328. https://doi.org/10.7717/peerj.328.

[8] Pimentel D. Species diversity and insect population outbreaks. Annals of Entomological Society of America. 1961; **54**: 76-86.

[9] Boero F, Belmonte G, Bussotti S, Fanelli G, Fraschetti S, Giangrande A, Gravili C, Guidetti P, Pati A, Piraino S, Rubino F, Saracino OD, Schmich J, Terlizzi A, Geraci S. From biodiversity and ecosystem functioning to the roots of ecological complexity. Ecological Complexity. 2004; **1**: 101-109.

[10] Ceballos G, García A, Ehrlich, PR. The sixth extinction crisis: loss of animal populations and species. Journal of Cosmology and Astroparticle Physics, 2010; **8**: 1821-1831.

[11] Dunn RR. Modern insect extinctions, the neglected majority. Conservation Biology. 2005; **19**: 1030-1036.

[12] Thomas JA, Telfer MG, Roy DB, Preston CD, Greenwood JJD, Asher J, Fox R, Clarke RT, Lawton JH. Comparative losses of British butterflies, birds, and plants and the global extinction crisis. Science. 2004; **303**: 1879-1881.

[13] Wagner D.L: Trends in biodiversity: Insects. In Encyclopedia of the Anthropocene Oxford, England: Elsevier. 2018. 131-143.

[14] Warren MS, Maes D, Swaay CAM, Goffart P, Dyck HV, Bourn NAD, Wynhoff I, Hoare D, Ellis S. The decline of butterflies in Europe: Problems, Significance and possible solutions. Proceedings of National Academy of Sciences of the United States of America. 2021; **118**(2):e200255117. https://doi.org/10.1073/pnas.2002551117.

[15] Schowalter TD, Pandey M, Persley SJ, Wilig MR, Zimmerman JK.

Arthopods are not declining but are responsive to disturbance in the Luquillo Experimental Forest, Puerto Rico. Proceedings of National Academy of Sciences of the United States of America. 2021; **118**(2): e2002556117. https://doi.org/10.1073/pnas.2002556117.

[16] Wagner DL, Fox R, Salcido DM, Dyer LA. A window to the world of global insect declines: Moth biodiversity trends are complex and heterogenous. Proceedings of National Academy of Sciences of the United States of America. 2021; **118**(2): e2002549117. https://doi.org/10.1073/pnas.2002549117.

[17] Boyes DH, Fox R, Shortfall CR, Wittaker R. Bucking the trend: The diversity of Anthropocene 'winners' among British moths. 2019; **11**(3): e43862. https://doi.org/10.21425/F5FBG43862.

[18] Klink RV, Bowler DE, Gongalsky KB, Swengel AB, Gentile A, Chase JM. Meta-analysis reveals declines in terrestrial but increases in freshwater insect abundances. Science. 2020; **368**(6489):417-420. https://doi.org/10.1126/science.aax9931.

[19] Graves SD, Shapiro AM. Exotics as host plants of the California butterfly fauna. Biological Conservation. 2003; **110**: 413-433.

[20] Deutsch C A, Tewksbury JJ, Huey RB, Sheldon KS, Ghalambor CK, Haak DC, Martin PR. Impacts of climate warming on terrestrial ectotherms across latitude. Proceedings of the National Academy of Sciences of the United States of America. 2008; **105**: 6668-6672.

[21] Drizo R, Young SH, Galetti M, Ceballos G, Isaac NJB, Collen B. Defaunation in the Anthropocene. Science. 2014; **345**: 401-406.

[22] Keil P, Storch D, Jetz W. On the Decline of Biodiversity Due to Area Loss. Nature Communications. 2015; **6**: 8837. Available online: https://doi.org/10.1038/ncomms9837.

[23] Bridgewater P, Loyau A, Schmeller DS. The seventh plenary of the governmental platform for biodiversity and ecosystem services (IPBES-7): a global assessment and a reshaping of IPBES. Biodiversity and Conservation. 2019; **28**: 2457-2461. Available online: https://doi.org/10.1007/s10531-019-01804-w

[24] Habitat loss. National Wildlife Federation. Available online: https://www.nwf.org/Educational-Resources/Wildlife-Guide/Threats-to-Wildlife/Habitat-Loss.

[25] Hanski I. Meta-population Ecology. Oxford University press, Oxford, United Kingdom. 1999; p 324.

[26] Habel JC, Schmitt T. Vanishing of the common species: Empty habitats and the role of genetic diversity. Biological Conservation. 2018; **218**: 211-216.

[27] Ceballos G, Ehrlich PK, Drizo R. Biological annihilation via the ongoing sixth mass extinction signaled by vertebrate population losses and declines. Proceedings of National Academy of Sciences of the United States of America. 2017; **114** (30): E6089-E6096. https://doi.org/10.1073/pnas.1704949114.

[28] Dudley N, Alexander S. Agriculture and biodiversity: a review. Biodiversity. 2017; **18**: 45-49.

[29] Mule R, Sabella G, Robba L, Manachini B. Systematic review of the effects of chemical insecticides on four common butterfly families. Frontiers in Environmental Science. 2017; **5**:32.

[30] Lundgren JG, Hesler LS, Clay SA, Fausti SF. Insect communities in soybeans of eastern South Dakota: the effects of vegetation management and pesticides on soybean aphids, bean leaf beetles and their natural enemies. Crop Protection. 2013; **43**: 104-118.

[31] Goulet H, Masner L. Impact of herbicides on the insect and spider diversity in eastern Canada. Biodiversity. 2017; **18**:50-57.

[32] Roessink I, Merga LB, Zweers HJ, Brink PJ. The neonicotinoids imidacloprid shows high chronic toxicity to mayfly lymphs. Environment Toxicology and Chemistry. 2013; **32**:1096-1100.

[33] Pecenka JRJ, Lundgren, G. Non-target effects of clothianidin on monarch butterflies. Naturwissenschaften. 2015; **102**: 19.

[34] Goulson D. An overview of the environmental risks posed by neonicotinoid insecticides. Journal of Applied Ecology. 2013; **50**: 977-987.

[35] Nakanishi K, Yokomizo H, Hayashi T. Were the sharp declines of dragonfly populations in the 1990s in Japan caused by fipronil ans imidacloprid? An analysis of Hill's causality for the case of sympetrum frequens. Environmental Science and Pollution Research. 2018; **25**: 35352-35364. Available online: https://doi.org/10.1007/s11356-018-3440-x.

[36] Hahn M, Schotthöfer A, Schmitz J, Franke LA, Brühl CA. The effects of agrochemicals on Lepidoptera, with a focus on moths, and their pollination service infield margin habitats. Agriculture, Ecosyststems and Environment. 2015; **207**: 153-162.

[37] Brittain CA, Vighi M, Bommarco R, Settele J, Potts SG. Impacts of a pesticide on pollinator species richness at different spatial scales. Basic and Applied Ecology. 2010; **11**: 106-115.

[38] Krischik V, Rogers M, Gupta G, Varshney A. Soil-applied imidacloprid translocates to ornamental flowers and reduces survival of adult *Coleomegilla maculata*, *Harmonia axyridis*, and *Hippodamia convergens* lady beetles, and larval *Danaus plexippus* and *Vanessa cardui* butterflies. PLoS One. 2015; **10**: e0119133.

[39] Jinguji H, Thuyet DQ, Uéda T, Watanabe H. Effect of imidacloprid and fipronil pesticide application on *Sympetrum infuscatum* (Libellulidae: Odonata) larvae and adults. Paddy Water Environment. 2013; **11**: 277-284. Available online: https://doi.org/10.1007/s10333-012-0317-3.

[40] Weston DP, Schlenk D, Riar N, Lydy MJ, Brooks ML. Effects of pyrethroid insecticides in urban runoff on Chinook salmon, steelhead trout, and their invertebrate prey. Environmental Toxicology and Chemistry. 2015; **34**: 649-657.

[41] Hallmann CA, Sorg M, Jongejans E, Siepel H, Hofland N, Schwan H, Stenmans W, Müller A, Sumser H, Hörren T, Goulson D, de Kroon H. More than 75 percent decline over 27 years in total flying insect biomass in protected areas. PLoSOne. 2017; **12**: e0185809.

[42] Wallis DeVries, MF, Swaay CAM. Global warming and excess nitrogen may induce butterfly decline by microclimatic cooling. Global Change Biology. 2006; **12**: 1620-1626.

[43] Klop E, Omon B, Wallis DeVries MF. Impact of nitrogen deposition on larval habitats: The case of the Wall Brown butterfly *Lasiommata megera*. Journal of Insect Conservation. 2015; **19**: 393-402.

[44] Stevens CJ, Dise NB, Mountford JO. Impact of nitrogen deposition on the species richness of grasslands. Science. 2004; **303**(5665): 1876-1879. Available online: https://doi.org/10.1126/science.1094678.

[45] Biesmeijer JC, Roberts SPM, Reemer M, Ohlemuller R, Edwards M, Peeters T, Schaffers AP, Potts SG, Kleukers R, Thomas CD, Settele J, Kunin WE. Parallel declines in pollinators and insect-pollinated plants in Britain and the Netherlands. Science. 2006; **313**: 351-354.

[46] Fox R. The decline of moths in Great Britain: a review of possible causes. Insect Conservation and Diversity. 2013; **6**: 5-19.

[47] Brehm G, Niermann J, Jaimes-Nino LM, Enseling D, Jüstel T, Axmacher JC, Fiedler K: Moths are strongly attracted to ultraviolet and blue radiation. Insect Conservation and Diversity. 2021; **14**: 188-198.

[48] Deichmann, JL, Ampudia Gatty, C, Andía Navarro, JM, Alonso A, Linares-Palomino R, Longcore T. Reducing the blue spectrum of artificial light at night minimizes insect attraction in a tropical lowland forest. Insect Conservation and Diversity. 2021; **14**: 247-259.

[49] Langevelde FV, Grunsven RH, Veenendaal E, Fijen T. Artificial lightining inhibits feeding in moths. Biology Letters. 2017; **13**(3): 20160874. Available online: https://doi.org/10.1098/rsbl.2016.0874.

[50] Boyes DH, Evans DM, Fox R, Parsons MS, Pocock MJO. Is light pollution driving moth population declines? A review of causal mechanisms across the life cycle. Insect Conservation and Diversity. 2021; **14**: 167-187.

[51] Kalinkat G, Grubisic M, Jechow A, van Grunsven RHA, Schroer S, Hölker F. Assessing long-term effects of artificial light at night on insects: what is missing and how to get there. Insect Conservation and Diversity. 2021; **14**: 260-270.

[52] Ewell JJ. "Deliberate introductions of species: Research needs -Benefits can be reaped, but risks are high". BioScience. 1999; **49** (8): 619-630. Available online: doi:10.2307/1313438. JSTOR 1313438.

[53] Rejmanek M, Richardson DM. What makes some conifers more invasive Proceedings of the Fourth International Conifer Conference. 2000.

[54] Sodhi NS, Brook BW, Brasshaw CJA. Causes and consequences of species extinctions. In Levi *et al.*, The Princeton Guide to Ecology. Princeton University Press-New Jersy. 2009; p 832.

[55] Hill GM, Kawahara AY, Daniels JC, Bateman CC, Scheffers BR. Climate change effects on animal ecology: butterflies and moths as a case study. Biological Reviews. 2021. Available online: https://doi.org/10.1111/brv.12746.

[56] Habel JC, Segerer AH, Ulrich W, Schmitt T. Succession matters: Community shifts in moths over three decades increases multi-functionality in intermediate successional stages. Scientific Reports. 2019; **9**: 5586.

[57] Yekwayo I, Pryke JS, Gaigher R, Samways MJ. Only multi-taxon studies show the full range of arthropod responses to fire. PLoSOne. 2018; **13**(4):e0195414.

[58] Macgregor CJ, Thomas CD, Roy DB, Beaumont MA, Bell JR, Brereton T, Bridle JR *et al*. Climate-induced phenology shifts linked to range expansions in species with multiple reproductive cycles per year. Nature Communications. 2019; **10**: 4455. Available online: https://doi.org/10.1038/s41467-019-12479-w.

[59] Melero Y, Stefanescu C, Pino J. General declines in Mediterranean butterflies over the last two decades are modulated by species traits. Biological Conservation. 2016; **201**:336-342.

[60] Dyck HV, Puls R, Bonte D, Gotthard K, Maes D. The lost generation hypothesis: Could climate change drive ectotherms into a developmental trap? Oikos. 2015; **124**: 54-61.

[61] Schowalter TD: Insect ecology: An ecosystem approach. 3rd ed. Academic Press, San Diego. 2011; p 633. Available online: https://doi.org/10.1016/c2009-0-60945-4.

[62] Kerr J, Pindar A, Galpern P, Packer L, Potts SG, Roberts SPM, Rasmont P, Schweiger O *et al*. Climate change impacts on bumble bees converge across continents. Science. 2015; **349** (6244):177-180.

[63] Pyke GH, Thomson JD, Inouye DW, Miller TJ. Effects of climate change on phenologies and distributions of bumble bees and the plants they visit. Ecosphere. 2016; 7:e01267.

[64] Wu CH, Holloway JD, Hill JK, Thomas CD, Chen IC, Ho CK. Reduced body sizes in climate-impacted Borneo moth assemblages are primarily explained by range shifts. Nature Communications. 2019; **10**: 4612.

[65] Canale CI, Henry PY. Adaptive phenotypic plasticity and resilience of vertebrates to increasing climatic unpredictability. Climate Research. 2010; **43**(1-2): 135-147.

[66] Kleckova I, Klecka J. Facing the heat: thermoregulation and behaviour of lowland species of a cold-dwelling butterfly genus, *Erebia*. PLoS One. 2016; **11**(3): 1-16.

[67] Watt WB. Adaptation at specific loci. II. Demographic and biochemical elements in the maintenance of the *Colias* pgi polymorphism. Genetics. 1983; **103**: 691-724.

[68] Niitepold K, Smith AD, Osborne JL, Reynolds DR, Carreck NL, Martin AP, Marden JH, Ovaskainen O, Hanski I. Flight metabolic rate and Pgi genotype influence butterfly dispersal rate in the field. Ecology. 2009; **90**(8): 2223-2232.

[69] Mitikka V, Hanski I. Pgi genotype influences flight metabolism at the expanding range margin of the European map butterfly. Annales Zoologici Fennici. 2010; **47**(1): 1-14.

[70] Watt WB. Eggs, enzymes, and evolution: natural genetic variants change insect fecundity (allozymes/fitness components/global warming/phosphoglucose isomerase/thermal ecology). Proceedings of the National Academy of Sciences of the United States of America. 1992; **89**(22): 10608-10612.

[71] Wang L, Yang S, Zhao K, Han L. Expression profiles of the heat shock protein 70 gene in response to heat stress in *Agrotis c-nigrum* (Lepidoptera: Noctuidae). Journal of Insect Science. 2015; **15**(1): 9.

[72] Travis JMJ. Climate change and habitat destruction: a deadly anthropogenic cocktail. Proceedings of the Royal Society of London: Biological Sciences. 2003; **270**: 467-473. Available online: https://doi.org/10.1098/rspb.2002.2246.

[73] Shepherd S, Lima MAP, Oliveira EE, Sharkh SM, Jackson CW, Newland PL. Extremely low frequency Electromagnetic fields impair the cognitive and motor abilities of honey bees. Science Reports. 2018; **8**(1): 1-9.

[74] Habel JC, Segerer A, Ulrich W, Torchyk O, Weisser WW, Schmitt T. Butterfly community shifts over 2 centuries. Conservation Biology. 2016; **30**: 754-762.

[75] Hewitt JE, Thrush SF, Ellingsen KE. The role of rare species identities in spatial patterns of species richness and conservation. Conservation Biology. 2016; **30**: 10180-11088.

[76] Ranjha MH, Irmler U. Movement of carabids from grassy strips to crop land in organic agriculture. Journal of Insect Conservation. 2014; **18**: 457-467.

[77] Woodcock BA, Bullock JM, Shore RF, Heard MS, Pereira MG *et al*. Country-specific effects of neonicotinoid pesticides on honey bees and wild bees. Science. 2017; **356**: 1393-1395. Available online: https://doi.org/10.1126/science.aaa1190.

[78] Samways MJ. Insect conservation for the twenty-first century. In: Shah M.M, Sharif U (eds) Insect science-diversity, conservation and nutrition. 2018; Intech Open. https://doi.org/10.5772/intech open.73864.

[79] White PJT, Kerr JT. Human impacts on environment–diversity relationships: evidence for biotic homogenization from butterfly species richness patterns. Global Ecology and Biogeography. 2007; **16**: 290-299.

Chapter 3

# Agricultural Intensification Causes Decline in Insect Biodiversity

*Mumuni Abudulai, Jerry Asalma Nboyine, Peter Quandahor, Ahmed Seidu and Fousséni Traore*

## Abstract

The world's population exceeded 7 billion in late 2011 and it is expected to reach 9.3 billion by 2050. Meanwhile, demand for food is predicted to increase between 50 and 100% by 2050. To meet the food demands of the increasing population, agricultural intensification practices including growing monocultures of high-yielding crop varieties and increased applications of fertilizers and pesticides have been used to increase productivity. These practices, however, impact negatively on biodiversity of existing flora and fauna, particularly causing huge declines in insect biodiversity. This chapter reviews present state of knowledge about agricultural intensification practices and global decline of insect biodiversity (i.e., pest and beneficial insect species) in intensive agricultural system and point out the likely drivers of these declines. It concludes the review by examining sustainable agricultural intensification practices that could be used to mitigate these biodiversity declines while maintaining productivity in intensive agricultural systems.

**Keywords:** insect decline, agricultural intensification, crop production, food demands, beneficial arthropods

## 1. Introduction

Global decline of biodiversity of many terrestrial and aquatic invertebrates, particularly insects, has been a major concern to biologists and ecologists. This is because biodiversity provides many important ecosystem services due their abundance and diversity [1–3]. Much of the decline has been blamed on human activities such as hunting, habitat loss through deforestation, agricultural expansion and intensification, industrialization and urbanization [4, 5], which together accounted for 30–50% encroachment on natural ecosystems at the end of the twentieth century [6]. Agricultural intensification is considered the key driver of this biodiversity loss in many taxa including birds, insectivorous mammals and insects. The removal of natural habitat elements such as hedgerows, trees and other landscape features together with the recurrent use of chemical fertilizers and pesticides in agricultural intensification systems negatively affect overall biodiversity of insects [7]. Extensive use of pesticides is reported as the primary factor responsible for the decline of birds in grasslands [8] and aquatic organisms in streams [9], with probably other factors contributing to or amplifying their effects.

Long-term population monitoring study at several protected areas of Germany revealed a 76% decline in flying insect biomass with an annual loss of 2.8% [10].

Similarly, a study in the rainforests of Puerto Rico showed biomass losses between 98 and 78% for ground-foraging and canopy-dwelling arthropods over a 36-year period, with annual losses between 2.7 and 2.2%, respectively [11]. The authors showed parallel declines in birds, frogs and lizards at the same areas, which they attributed to invertebrate food shortages. The studies above (10–11) confirm the declining trend in flying insects (mainly Diptera) reported earlier for parts of Southern Britain [12]. While climate change may be a contributory factor to arthropod declines, intensification practices including deforestation were reported to be responsible for the annual loss of insect biomass in the tropical rainforest of Germany [10]. The authors also pointed to the effect of synthetic pesticides as a likely driver of the losses in insect biomass.

The above studies demonstrate general knowledge about biodiversity decline in insects. It appears that insect declines are substantially greater than those observed in birds or plants [13], and this could have far reaching consequences on several of the world's ecosystems. This review summarizes current knowledge about insect declines; that is, the changes in species richness (biodiversity) and population abundance through time in intensive agricultural systems point to the likely drivers of these declines and conclude with management practices that could mitigate these declines in sustainable agricultural systems. Previous reviews are limited in scope to one or a few insect taxa (e.g., butterflies, carabids) in specific regions and made no comparisons across taxa in different geographical regions (e.g., Sequera et al. 2014; Zhao et al. 2015).

## 2. Agricultural intensification production practices

Agricultural production has struggled over the past few centuries to keep pace with the ever-increasing world population of humans, which exceeded 7 billion in late 2011 and is expected to reach 9.3 billion by 2050 [14]. In sub-Saharan Africa, for example, the current population of 1.1 billion people is projected to double over the next 30 years [15]. The increasing population increased demand for food and also brought in its wake increased demand for land for housing, roads and other infrastructural needs, which limited land availability for other purposes including agriculture. Thus, the hitherto traditional agricultural practices such as low-input agriculture with inherent low yields and shifting cultivation appeared no longer tenable in the quest to produce enough food for the growing population. This led to the intensification of agricultural practices more especially after World War II. In Europe and North America, the intensification of agriculture began in the first half of the twentieth century, whereas in South America, Africa and Asia, it started mainly in the second half of the century [16].

The agricultural intensification practices include expansion of farms into large commercial enterprises, accompanied by a changed emphasis to monocultures, and the application of increasing inputs of fertilizers and synthetic pesticides [14, 17]. Today's farmlands are larger in scope than their predecessors, more of monocultures, and more rely more on external inputs such as fertilizer, insecticide, and herbicide. In such systems, there is also greater emphasis on the elimination of weeds, cutting down hedgerows and trees in order to facilitate mechanization of fields. Surface waterways are also modified including stream channelization to ease flow and improve irrigation and drainage of fields. These intensification practices drastically reduce the level of refugia available for insects, herbaceous plants, vertebrate insectivores, and other organisms and consequently an overall decline in biodiversity, both in species numbers and in biomass [14, 18, 19].

More than a quarter of the world tropical forests have been cut since the ratification of the Convention on Biological Diversity in 1992, leaving many to wonder whether there will be any substantial stands of tropical forest remaining by the end of this century. Many grasslands and forested areas have also been converted into croplands and plantations [17]. The effects of these practices on biodiversity loss is further exacerbated by the effect climate change, which limit the location of favored regions for crops and other life forms [17].

## 3. Effects of agricultural intensification practices on arthropod biodiversity decline

A lot has been reported about the effects of agricultural intensification practices on biodiversity loss in insects. Zabel et al. [19] discussed the trade-offs between increasing agricultural intensification and biodiversity decline. Inevitably, increased structural modification of habitats and change in the heterogeneity of farmlands in agricultural intensification systems affect biodiversity. The intensive practices alter the availability of food and shelter for insects and other life forms, which affect the abundance and diversity of species (14). Consequently, major insect declines were observed when agricultural practices shifted from the hitherto low-input traditional farming to the intensive, industrial-scale production brought about by the Green Revolution [19]. In its wake, rare species associated with protected ecosystems and natural habitats retreated or were lost completely [18, 19]. Monocultures led to a great simplification of insect biodiversity among pollinators, insect natural enemies and nutrient recyclers, and created the suitable conditions for agricultural pests to flourish. Thus, agricultural intensification serves as the main driver of insect declines in both terrestrial and aquatic ecosystems [20–22].

Raven and Wagner [17] reported of increased clearing of forests in the tropics for crops, pasture and wood fuel in Central Africa, Central America, many parts of South America and Southeast Asia. An average of 5 million acres of the forest was lost annually to industrial-scale agriculture between 2001 and 2015 [23, 24]. This huge deforestation poses serious threats to the world's insect biodiversity as the majority of insect species diversity is found in the tropics. Deforestation is one of the major drivers of biodiversity loss and insect declines [17–25]. Moreover, deforestation on larger scales has the potential to change weather and rainfall patterns that may further impact negatively on insect populations [24, 25]. Insect biodiversity is very important for successful agriculture in providing many ecosystem services such as pollination, nutrient recycling and biological pest control.

In [26, 27], it was reported that agriculture is the primary contributing factor in insect losses in California and Ohio. According to [27, 28], butterfly diversity in southwest Germany began declining two centuries ago, but with steeper rate of declines observed after World War II, when intensification practices increased. Over the past half century, two-thirds of the common moth species in Great Britain are decreasing in number. Powney et al. [29] analyzed the long-term abundance trends of moths in Great Britain and reported that moth abundances had decreased by 31% over the past five decades [19]. Similarly, in [17], the elevated rate of loss was reported in diverse group of insect fauna of the grassland world, including butterflies and noctuid moths (Lepidoptera); ants, bees and wasps (Hymenoptera); scarab and ground beetles (Coleoptera); crickets, grasshoppers and katydids (Orthoptera); leaf and plant hoppers, seed bugs and their kin (Heteroptera). Further, there are reports of declines of wild bees, particularly from northwestern Europe due to agricultural intensification [30].

## 4. Biodiversity declines of selected insect groups

This part of the chapter discusses in detail biodiversity declines of selected insect groups caused by the effect of agricultural intensification practices across the globe.

### 4.1 Lepidoptera

Butterflies and moths have high level of host plant specialization and are therefore vulnerable to habitat deterioration [31]. They also have a wide range of distribution and important for the delivery of key ecosystem services such as biological pest control and pollination [32]. Moths, which are about 10 times more different than butterflies, constitute important prey items for bats and sustain population levels of a myriad of other insectivorous animals [33].

Maes and Van Dyck [34] pioneered report of drastic changes in butterfly biodiversity in Flanders (Belgium) during the twentieth century. They observed that habitat loss due to urbanization and agricultural intensification expansion resulted in a steady decline of 69% of 45 extant species [34]. A follow-up study in the Netherlands by van Dyck [35] also found that 11 out of 20 most common and widespread butterfly species declined in both distribution and abundance between 1992 and 2007. Moreover, local populations of *Lasiommata megera* and *Gonepteryx rhamni* are now endangered and two other species (*Aglais io* and *Thymelicus lineola*) are vulnerable. A parallel study in the Netherland of the range of distribution of 733 species of day-flying moths between 1980 and 2000 showed decline in 85% of species, with 38% critically endangered, 34% vulnerable and 15% threatened [36]. Similarly, a long-term survey at the Kullaberg Nature Reserve in Sweden showed that out of 269 species, 45% declined, 22 were threatened and 159 species were extinct [37]. Monophagous and oligophagous species feeding on grass or herbs in wetlands declined more than those feeding on deciduous trees and shrubs. Also, historical records of 74 butterfly species in Finland showed that 60% of grassland species declined, whereas 86% of generalist species and 56% of those living at forest edge ecotones increased in abundance [38]. For the same locati0n, monitoring the population of 306 noctuid moth species showed drastic declines for species with comparatively longer flight periods and smaller geographical range [39]. Similar findings were reported for northeastern Spain, where in-depth study of the population trend for 66 butterfly species showed a decline in 46 species, while 15 species had increased in abundance and 5 remained stable [40]. A comprehensive report on the status of 576 species of butterflies in Europe found that 71 were threatened and declined over a 25-year period [41]. The greatest declines were observed among specialist butterflies of grassland biotopes (19% species), wetlands and bogs (15%) and woodlands/forests (14%), due to habitat conversion into croplands and intensification of agricultural practices; pesticides negatively affected 80% species. Some species (*Lopinga achine* and *Parnassius apollo*) had declined due to afforestation, that is, conversion of open woodland habitats to dense forests. A recent assessment of 435 butterflies species native to Europe revealed that 19% of the species are declining, while 8.5% species are threatened, and three endangered, *viz. Pieris brassicae wollastoni*, *Triphysa phryne* and *Pseudochazara cingovskii* [42]. A comprehensive database from the UK showed that 41 out of 54 common butterflies species had been declining since the 1970s, with 26% of species showing decreases over 40% of their range [30]. The authors suggested habitat fragmentation and/or destruction and intensification of agriculture, including the increased usage of chemical fertilizer and pesticides, as the possible drivers for this biodiversity loss.

Long-term monitoring data of butterflies are limited in the United States. However, surveys in prairie habitats and bogs of Wisconsin and Iowa showed fluctuating populations of certain species. These fluctuations were driven by habitat modification and moisture levels dependent on climate change [43]. Surveys of 67 butterfly species in California between 1972 and 2012 showed initially stable populations until 1997 when populations dropped steeply to 23 species. The observed declines correlated significantly with percentage of land converted to agriculture and usage of insecticide, with neonicotinoid being the most important. The declining trend in 1997 followed the introduction of the neonicotinoid insecticides in that State [44]. In Massachusetts, the distributional ranges of 116 species shifted northward between 1992 and 2010. Two southern species adapted to warmer conditions expanded in range (*Papilio cresphontes* and *Poanes zabulon*), while populations of 80% of butterflies declined in southern parts of that State [45]. Although survey records are limited, Lepidoptera declines appear to be less dramatic in certain parts of the Asian region. In Japan, 15% of 240 species of butterflies are threatened, with 80% of grassland species being endangered, and two species (*Melitaea scotosia* and *Argynnis nerippe*) close to extinction in the national territory [46]. The steady intensification of Japan's traditional "satoyama" landscape (i.e., a mosaic of rice paddy fields, grassland and coppice forests) has negatively affected most species. In Malaysia, some 19% of moths at Mount Kinabalu (Borneo) had their abundance reduced between 1965 and 2007 (**Table 1**) [47].

### 4.2 Hymenoptera

Members of this group include bees, ants and wasps. They provide many important ecosystem services such as pollination and biological control of insect pests. Bees are essential pollinators of flowering plants and constitute a third of all pollinators [19]. Also, honey bees have been managed for millennia as a source of honey and beeswax. Hence, a need for information about their population status because of the important ecosystem services they provide [55].

A report on 18 bee species in Britain showed declining trends for seven species since the 1960s. The species with the most declines were *Bombus humilis, B. ruderatus, B. subterraneus* and *B. sylvarum*) [56]. The declines were associated with extensive use of chemical fertilizers as a source of nitrogen [57]. In Denmark, five of 12 native species were extinct, whereas the once common *Bombus distinguendus*

| Insect taxon | Declining (%) | Threatened (%) | Extinction rate (%) | Reference |
|---|---|---|---|---|
| Coleoptera | 49 | 34 | 6.6 | [48] |
| Diptera | 25 | 0.7 | n.a | [49] |
| Ephemeroptera | 37 | 27 | 27 | [50] |
| Hemiptera | 8 | n.a | n.a | [51] |
| Hymenoptera | 46 | 44 | 15 | [52] |
| Lepidoptera | 53 | 34 | 11 | [13] |
| Odonata | 37 | 13 | 6 | [53] |
| Orthoptera | 49 | n.a | n.a | [1] |
| Plecoptera | 35 | 29 | 19 | [54] |
| Trichoptera | 68 | 63 | 6.8 | [49] |

**Table 1.**
*Proportion of declining and threatened species per taxa according to IUCN criteria.*

is currently classified as a threatened species [58]. In central Europe, 48 out of 60 species and subspecies have declined over the past 136 years. Of this, 30% are considered as endangered species, while four are extinct [59]. These extinctions are associated with agricultural intensification initiated by the Green Revolution in the second half of the twentieth century [59]. Pollinator declines were reported in Swedish red clover fields since 1940 with only two rare species (*B. terrestris* and *B. lapidaries*) remaining stable [59]. Similar to Denmark, *B. distinguendus* is extinct in the southern part of Sweden, with agricultural expansion and extensive use of agrochemicals reported as the major drivers for biodiversity decline in bees observed over the past 75 years [60]. Similar declining trends were observed among 46% of the *Bombus* species in Europe, of which 24% are endangered and one species (i.e., *B. callumanns*) showing >80% decline due to extensive application of chemical fertilizers in agricultural areas. Further, studies in North America showed that 50% of the 14 bumblebee species in southern Ontario (Canada) were declining. However, three species (*B. bimaculatus, B. impatients* and *B. rufocinctus*) were increasing in abundance, while another three (*B. affinis, B. pensylvanicus* and *B. terricola*) were extinct [61, 62]. In the midwestern regions, a survey on 16 species of bumblebees showed a 50% population decline, while four species (*B. borealis, B. ternarius, B. terricola* and *B. variabilis*) were extinct [18]. A similar decline trend was observed at Itasca State Park (Minnesota), where 11 out of 30 species of stingless bees (Megachilidae) declined, whereas 11 were missing [63–65] due to herbicide use and agricultural intensification.. On a national scale, where historical records were compared with intensive surveys across 382 locations in the USA, 50% of the initial 96% population declined in the last 30 years, and their habitat was condensed to between 23 and 87% [66]. Also in the USA, 3.5 million out of 6.0 million honey bee colonies reported declines, over the past six decades, representing 0.9% loss per year [67]. These declines were linked to the use of dichloro-diphenyl-trichloroethane (DDT) in agriculture [68], toxic pesticides for the management of Varroa mites [69, 70] and poor nutritional value of agro-landscapes dominated by monocultures (e.g., corn, oilseed rape, cotton) [71]. Declines have been reported for bees in Brazil (63%), Costa Rica (60%) and Finland (23%) [72, 73]. Again, these declines were attributed habitat loss due to agricultural intensification practices [74–77]. Other factors contributing to the loss of bees are colony collapse disorder (CCD) caused by pathogens, toxins, parasites and other stressors [58, 78]. Presently, about 40, 30, 29 and 3–13% of colonies are lost annually in USA, Europe, South Africa and China, respectively [55]. The use of pesticides containing neonicotinoids and fipronil is implicated in these losses [58, 78, 79]. These pesticides inhibit the reproductive performances of queens and drones [80, 81], thus compromising the long-term viability of entire colonies [82].

In general, studies [83–85] identified four major phases of bee extinction particularly in Britain. These are as follows: i) the second half of the nineteenth century, with the introduction of guano fertilizers and conversion of arable crops to permanent grasslands, which reduced floral resources; ii) after the First World War, when florally-diverse crop rotations were replaced with chemical fertilizers; iii) between 1930 and 1960, when most species went extinct probably due to changes in agricultural policy (i.e., Green Revolution) that fostered agricultural intensification; and iv) from 1987 to 1994, when rates of decline slowed down perhaps because the most sensitive species were already lost or reduced substantially [20].

Apart from bees, the status of most other hymenopterans (i.e., ants, wasps and parasitoids) that provide important ecosystem services remains practically unknown to date (**Table 2**). There is, therefore, a need for intensive research on these species.

| Taxon | Abundance | Decline | Location | Reference |
|---|---|---|---|---|
| **Hymenoptera** | | | | |
| Bumble bees | 18 species | 7 species | England | [56] |
| Bumble bees | 14 species | 8 species | Canada | [61] |
| Bumble bees | 60 species | 48 species | Central Europe | [59] |
| Honey bees | 6-m colonies | 3.5-m colonies | USA | [67] |
| Wild bees | | 52% population | Britain | [85] |
| Wild bees | | 67% population | Netherlands | [63] |
| Wild bees | | 32% population | North America | [64] |
| Cuckoo wasps | | 23% population | Finland | [85] |
| Stingless bees | 30 species | 11 species | USA | [64] |
| Orchid bees | 24 species | 64% species | Brazil | [72] |
| Parasitic wasps | 48 species | 23% species | Finland | [85] |
| **Coleoptera** | | | | |
| Ground beetles | 419 species | 34% species | Belgium, Denmark | [86] |
| Ground beetles | 49 species | 16% species | UK | [87] |
| Ladybird beetles | | 68% species | USA | [86] |
| Dung beetles | | 31% population | Italy | [88] |
| Saproxylic beetles | 436 species | 57% species | Europe | [89] |
| **Odonata** | | | | |
| Dragonflies | 52 species | 65% population | USA | [90] |
| Odonata species | 200 species | 57 species | Japan | [91] |
| Odonata species | 155 species | 13 species | South Africa | [92] |
| **Plecoptera** | | | | |
| Stoneflies | 14 species | 5 species | Czech Republic | [93] |
| Stoneflies | 77 species | 29% species | USA | [94] |
| **Ephemeroptera** | | | | |
| Mayflies | 107 species | 43% species | Czech Republic | [95] |

**Table 2.**
*Status of some taxa and their geographical areas.*

### 4.3 Diptera

Hoverflies (Syrphidae) are not only important pollinators, but vital biological pest control agent for pest such as aphids, with a preference for damp habitats. Most studies in the Mediterranean countries showed significant differences in diversity within this taxon, with 249 species alone in Greece [77] and 429 in Spain [96]. This notwithstanding, the only long-term study to date shows reductions in species richness among hoverflies in the Netherlands and the U.K. [76].

### 4.4 Coleoptera

This insect group contributes greatly to ecosystem management through control of pests and decomposition of organic matter [97]. Habitat destruction, extensive application of toxic chemicals and urbanization are the main causes of their decline.

Of 419 species surveyed in the Netherlands, Belgium, Luxemburg and Denmark, there was a 34% decline for carabids, with over 50% of xerophilic species of the genera *Amara*, *Harpalus* and *Cymindis* and *Carabus* recording a decrease in their populations [98]. Populations of those with large mobility potential remained stable [87]. In the U.K., among 68 carabids at 11 geographical areas, 49 declined, with 26 species considered susceptible and eight threatened, although populations of 19 species remained stable. Generally, 16% of the species were considered extinct throughout the 15-year period of study [99]. There was a 64% species decline in mountainous regions, 31% in moorlands and 28% in pastures. These declines were linked to microclimatic changes and habitat destruction [99]. In a study in New Zealand, 12 species of large carabid beetles were threatened, while another 36 declined with the two genera *Mecodema* and *Megadromus* being the most affected [100].

A study of 62 historical datasets of ladybird species in the USA and Canada showed stable species richness and population abundance [86]. However, a 68% decline was observed over a 20-year period in 1986 [86]. Two local species (i.e., *Adalia bipunctata* and *Coccinella novemnotata*) were classified as extinct in the north-eastern USA [101]. In addition to agricultural intensification and habitat change, competitive displacement by foreign generalist species, such as *C. septempunctata* and *Harmonia axyridis* [102], were identified as potential causes of the decline [103–106]. In the Czech Republic, populations of six species declined, while seven others increased out of 13 species studied [107].

Studies on the trends of dung beetle abundance and distribution are obtainable only for the Mediterranean region, which has the largest range of dung beetles in Europe [108]. A study in Spain indicated that out of the 55 native species, nine had declined from 28 to 7% loss, while their distributional range contracted from 48 to 29% [108]. *Scarabaeus pius* and *Gymnopleurus mopsus* were the most threatened species. Multivariate analyses showed that commercial farming, urbanization, and extensive use of pesticides were responsible for the declines [108]. Further, a study in Italy showed 31% decline in roller dung beetles and nine were extinct [88]. The trend of decline commenced from 1960s (two species), increased in the 1970s (three species), and peaked in the 1980s (six species). The possible primary decline factors were conversion of pastures to forests, agricultural intensification and a shift from free-ranging to stalled livestock management that reduced dung availability to foraging beetles.

Studies of scarab beetles showed that two *Scarabaeus* and four *Gymnopleurus* species are threatened, while *G. mopsus* is probably extinct [109, 110]. In France, a survey of the coastal region of France in 1996 showed nine Scarabaeidae were threatened and two Aphidiidae declined while Geotrupidae were extinct [111]. An earlier study showed 45-fold decline of *Scarabaeus semipunctatus* [89, 112]. In Europe, a study on saproxylic beetles showed that deforestation, agricultural intensification and wood harvesting caused destruction of native forests, thus endangering the survival of 56 beetle species. The two species, *Glaphyra bassetti* (Cerambycinae) and *Propomacruscypriacus* (Euchiridae), were the most threatened [109]. Since the abundance and distribution of 57% of the 436 known species are unidentified, the number of declining species could be even higher [113].

## 4.5 Hemiptera

These are distinctive phytophagous insects of plane regions, associated with natural and anthropogenic grasslands areas [85]. Sweep-net samples collected from 1963 to 1967 were compared with those collected from 2008 to 2010 at the same sites regarding species diversity, species composition and abundance. Generally, there was no change in species richness, irrespective of the strong interannual variability in abundance and weather condition. However, a decline in 14 species was observed while there was an increase in nine others and one species (*Zyginidia scutellaris*) increased in abundance and distribution. Median abundance decreased by 66% (from 679 to 231 individuals per site) over the 47-year period [114, 115]. The primary cause factors were attributed to airborne and soil acidification, partly due to agricultural intensification.

### 4.6 Orthoptera

A wide-ranging study on grasshoppers and crickets was conducted in Germany [116]. There was no fluctuation in their biodiversity and abundance over four decades (median nine species per site), and variations in species groupings were small. The only significant change was a steep decline in the Grasshopper of bare soils, *Myrmeleotettix maculatus*, while there was an increase in two generalist cricket species, *Tettigonia viridissima* and *Phaneroptera falcate*. Contrasting with other taxa, few Orthoptera species exhibited noticeable decline trends, possibly because most species are highly adaptable polyphagous grazers. Nonetheless, about half of the species are considered threatened in Germany.

### 4.7 Odonata

Dragonflies (Anisoptera) and damselflies (Zygoptera) are a small group of insects that contribute to the management of nuisance mosquitoes and agricultural pests [117, 118]. Of 118 aquatic insect species that are threatened, 106 are from the order Odonata [94, 119]. A study of 42 sites across USA recorded a decline in 52 species of dragonflies and damselflies, while there was an increase in 29 species over the 98-year period. Nine pollution-tolerant species declined significantly, including four species (*Sympetrum danae*, *Sympetrum costiferum*, *Ophiogomphus occidentis* and *Libellula nodisticta*) that were in an earlier survey [90]. In Europe, 15% of 138 Odonata species are currently endangered, with two damselflies (*Ceriagrion georgifreyi* and *Pyrrhosoma elisabethae*) and one dragonfly (*Cordulegaster helladica* sp. kastalia) highly threatened with extinction. Major declines for these insects occurred through the post-1960 agricultural intensification, with pollution of irrigation water by urban runoff and extensive application of agrochemicals being major causes [120]. In Japan, 57 out of 200 Odonata species are declining while 19 are threatened [91]. The greatest losses of populations are among lentic species once common in rice paddy fields, with the red dragonflies (*Sympetrum* spp.) experiencing the sharpest decline since the mid-1990s [121]. This decline has been associated with the use of fipronil and neonicotinoid insecticides [122, 123]. Similarly, of the 155 Odonata species recorded in South Africa, 13 are declining, while four others are extinct [92]. The authors opined that fortification of rare species in natural reserves of the country does not guarantee their survival, as current livestock management and other human activities negatively impact on their population.

### 4.8 Freshwater taxa

Freshwater insect taxa mostly exhibit inflexible life cycles, with several species being univoltine, thus making them vulnerable to habitat modifications. Flow changes, habitat fragmentation, pollution and invasive species are the main threats to all aquatic organisms, including insects [124, 125]. Data for three main orders of freshwater insects, Plecoptera, Ephemeroptera and Trichoptera, are reported here. There were no records found for Coleoptera (e.g., Dytiscidae, Hydrophilidae),

Hemiptera (e.g., Notonectidae, Gerridae) or Diptera (e.g., Chironomidae, Tipulidae).

### 4.8.1 Plecoptera

Stoneflies are ecologically important and characterized by high degrees of endemism and narrow ecological requirements [126]. Previous report in Europe showed a disappearance of *Aeniopteryx araneoides* and *Oemopteryx loewi* over the entire continent, while *Isogenus nubecula* was locally extinct [127]. The level of extinction ranges from 50% in Switzerland to 13–16% in the Mediterranean countries such as Spain and Italy. Up to 63% of the 516 European species of stoneflies are vulnerable to habitat destruction and climate change [128]. Stoneflies are susceptible to variation in water flow, even though they show resistance to acidification as compared with other macro-invertebrates [129]. A study of 78 stonefly species at 170 sites in the Czech Republic reported that low- and mid-altitude streams accounted for three quarters of the changes in species diversity. This was mainly due to pollution, impoundment and channelization at those sites [130]. Lowland river habitats indicated five endangered species of the 14 native species documented in the nineteenth century, while four were considered extinct. Over a 50-year period, 12% of the species disappeared, whereas two new species (*Brachyptera monilicornis* and *Leuctra geniculata*) appeared. Moreover, 22% of species reported had declined by >50%, including common species such as *Perla abdominalis, Amphinemura standfussi* and *Nemurella pictetii,* while a further 10% had become vulnerable.

Unlike terrestrial taxa, most declines were found among habitat generalists and less in specialized species (60–70%), which are tolerant to organic pollution. Sites affected by organic pollution showed only 17–33% decline of subtle and eurytopic species since the mid-1990s [93]; certain amount of species recovery has been detected following pollution modification in acidified habitats [131]. In Switzerland, 50% of the species of stoneflies and mayflies were lost between 1940s and 1980s [132], and similar trend occurred in other European countries and the USA, where 29% of the 77 local stonefly species were lost and 62% of the remainder became endangered over the past century [94]. Main losses occurred in the large rivers and agricultural areas during the 1940s and 1950s, when both agricultural intensification and urban expansion took place. Modification of river flows, channels and drainages structures was considered the driving factors for the declines. The large, long-lived species of Perlidae (summer stones) and Perlodidae (spring stones) were impacted the most, and 36% of summer stones had gone extinct since 1860. For sensitive genera such as *Acroneuria*, 88% of populations in the entire contingent were lost over the past century, whereas genera tolerant to organic pollution such as *Perlesta* had increased fourfold.

### 4.8.2 Ephemeroptera

A checklist of mayfly species in the Czech Republic identified 107 species of which four are considered extinct, seven critically threatened, another seven endangered, 16 vulnerable and 14 near threatened [95]. A comparison of local mayfly also showed variations in species abundance, distribution and composition, but no major declines were observed in biodiversity except for the large lowland rivers, which lost five specialist species [133]. Biodiversity improved slightly in the mid- and upper streams and rivers, possibly because of substantial reduction in water pollution post-1989 [93]. Two species became extinct (*Isonychia ignota* and *Ephemerella mesoleuca*), three became very rare, 11 were declining and nine were expanding their range, including the dominant *Centroptilum luteolum* and *Baetis niger* [93]. Main variations were due to losses of previously common and widespread species such as *B. alpinus* and *Epeorus assimilis*. The general difference among sites (15–30%) was mainly driven by species replacement. The present communities have shifted toward more simplified and less specialized assemblages in large rivers, whereas mayflies in small creeks have been replaced with species tolerant to pollution and siltation [93, 133]. In North America, a total of 672 species of mayflies are listed though no details are available about their abundance and distribution [134]. A collection for North and South Carolina (USA) reported 204 species [135], but again no status was indicated. A later study in relation to 10 rare species revealed, however, that four of the species sampled in the early twentieth century should be considered extinct [136].

#### *4.8.3 Trichoptera*

A comprehensive study on caddiflies species recorded 278 species in comparatively uninterrupted regions of Minnesota (USA) since the 1890s. Among the 278 species, 6–37% have declined in different areas, especially within the Limnephilidae (44% of species), Phryganeidae (21%) and Leptoceridae (12%) families [137]. *Agrypnia glacialis* and *Anabolia sordida* are presently considered extinct, whereas 17 rare species are yet to be found since the 1950s [137]. Entirely affected species are either in the univoltine or in the semivoltine family and because of their long life span and feeding habits, are mainly susceptible to anthropogenic disturbances in water courses. The majority of losses are found among shredder (72%) and predatory species (11%), which agrees well with losses of aquatic taxa in other countries [138, 139].

## 5. Sustainable agricultural intensification practices to mitigate biodiversity declines

The reports above show clearly that although agricultural intensification practices improve yields, they also impact negatively on the environment as evidenced by the decline in insect biodiversity. Biodiversity is an integral part of the natural resource base for agricultural production and therefore must be protected to sustain and safeguard the increased yields for the present and future generations. Over time, plant productivity decreases as biodiversity is lost [14]. A large proportion of studies (49.7%) point to habitat change as the main driver of insect declines, a factor equally implicated in global bird and mammal declines [135, 136]. Thus, habitat management practices are a key for sustainability of agricultural intensification practices. According to [136, 137], sustainable agricultural intensification is the management and conservation of the natural resource base and the orientation of technological and institutional change to ensure the attainment and continued satisfaction of human needs for present and future generations. It involves a process to produce high yields for existing land resource without affecting the environment. Sustainable intensification must include natural resource management practices that maintain the diversity of habitats as an intrinsic part of the agro-ecosystem or as additional land use interspersed among the fields (Firbank et al. 2008). These practices include crop rotation, reduced tillage, soil and water conservation, application of organic manure, intercropping and agroforestry [136, 138]. The practices will among other benefits ensure sustainable soil fertility through improved soil structure and soil microbial activities. Thus, sustainability requires the integration of multiple practices on a long-term basis to achieve desired environmental and agricultural outcomes. Intercropping with improved cultivars as well as integration of mixed crops with agroforestry and livestock could promote sustainable intensification and food security [137]. Also, the judicious implementation of integrated pest management (IPM) will minimize the use of toxic pesticides and enhance environmental safety for sustainable crop production [138]. Furthermore, in many of the world's farming systems, biological control constitutes an under-utilized and yet cost-effective tactic for pest control. The effect of biological control will be felt more in sustainable intensification systems such as those that involve IPM practices that are benign to natural enemies of pests and/or conserve biodiversity [139]. For aquatic insects, rehabilitation of marshlands and improved water quality are essential for biodiversity conservation and enhancement [140]. This may require the implementation of effective remediation technologies to clean the existing polluted waters [141, 142].

## 6. Conclusions

This chapter has provided a comprehensive discussion of effects of agricultural intensification on decline in insect biodiversity. Intensification practices highlighted as causes of this decline include expansion of farms into large commercial enterprises, cutting down hedgerows and trees in order to facilitate mechanization of farms, changed emphasis to monocultures, and increasing application of external input of fertilizers and synthetic pesticides. These practices largely reduce the level of refugia available for insects, herbaceous plants, vertebrate insectivores, and other organisms and consequently an overall decline in biodiversity, both in species numbers and in biomass. Insect biodiversity is integral to the resource base of the plant ecosystem that provides essential services for increased crop productivity and, therefore, must be protected to safeguard the survival of the present and future generations. To mitigate this decline therefore, the chapter highlights sustainable intensification practices to include habitat restoration practices such as intercropping, crop rotation, reduced tillage, agroforestry, application of organic manures coupled with drastic reductions in application of synthetic pesticides.

## Acknowledgements

The authors gratefully acknowledge management and staff of the Entomology Section of CSIR – SARI for their support in diverse ways during the preparation of this book chapter.

## Conflict of interest

The authors declare no conflict of interest in the publication of this book chapter.

## Author details

Mumuni Abudulai[1*], Jerry Asalma Nboyine[1], Peter Quandahor[1], Ahmed Seidu[1] and Fousséni Traore[2]

1 CSIR-Savanna Agricultural Research Institute, Tamale, Ghana

2 Laboratoire Central d'Entomologie Agricole de Kamboinsé (LCEAK), Institut de l'Environnement et de Recherches Agricoles (INERA), Ouagadougou, Burkina Faso

*Address all correspondence to: mabudulai@yahoo.com

## References

[1] Ceballos G, Ehrlich PR. Mammal population losses and the extinction crisis. Science. 2002;**296**:904-907

[2] Pimm SL, Raven P. Extinction by numbers. Nature. 2000;**403**(6772): 843-845

[3] Diamond JM. The present, past and future of human-caused extinctions. Philosophical Transactions of the Royal Society of London. B, Biological Sciences. 1989;**325**(1228):469-477

[4] Ceballos G, Ehrlich PR, Dirzo R. Biological annihilation via the ongoing sixth mass extinction signaled by vertebrate population losses and declines. Proceedings of the National Academy of Sciences. 2017;**114**(30):E6089-E6096

[5] Maxwell SL, Fuller RA, Brooks TM, Watson JE. Biodiversity: The ravages of guns, nets and bulldozers. Nature News. 2016;**536**(7615):143

[6] Vitousek PM, Mooney HA, Lubchenco J, Melillo JM. Human domination of Earth's ecosystems. Science. 1997;**277**(5325):494-499

[7] Tilman D, Fargione J, Wolff B, D'Antonio C, Dobson A, Howarth R, et al. Forecasting agriculturally driven global environmental change. Science. 2001;**292**:281-284

[8] Mineau P, Whiteside M. Pesticide acute toxicity is a better correlate of U.S. grassland bird declines than agricultural intensification. PLoS One. 2013;**8**:e57457

[9] Beketov MA, Kefford BJ, Schäfer RB, Liess M. Pesticides reduce regional biodiversity of stream invertebrates. Proceedings of the National Academy of Sciences. 2013;**110**(27):11039-11043

[10] Hallmann CA, Sorg M, Jongejans E, Siepel H, Hofland N, Schwan H, et al. More than 75 percent decline over 27 years in total flying insect biomass in protected areas. PLoS One. 2017;**12**:e0185809

[11] Lister BC, Garcia A. Climate-driven declines in arthropod abundance restructure a rainforest food web. Proceedings of the National Academy of Sciences. 2018. DOI: 10.1073/pnas.1722477115. (in press)

[12] Shortall CR, Moore A, Smith E, Hall MJ, Woiwod IP, Harrington R. Longterm changes in the abundance of flying insects. Insect Conservation and Diversity. 2009;**2**:251-260

[13] Thomas JA, Telfer MG, Roy DB, Preston CD, Greenwood JJD, Asher J, et al. Comparative losses of British butterflies, birds, and plants and the global extinction crisis. Science. 2004;**303**:1879-1881

[14] Emmerson M, Morales MB, Oñate JJ, Batary P, Berendse F, Liira J, et al. How agricultural intensification affects biodiversity and ecosystem services. Advances in Ecological Research. 2016;**55**:43-97

[15] Population Reference Bureau (PRB). World Population Data Sheet. 2019. Available from: https://interactives.prb.org/2020-wpds/ [Accessed date: 7 February 2020]

[16] Foley JA, DeFries R, Asner GP, Barford C, Bonan G, Carpenter SR, et al. Global consequences of land use. Science. 2005;**309**(5734):570-574

[17] Raven PH, Wagner DL. Agricultural intensification and climate change are rapidly decreasing insect biodiversity. Proceedings of the National Academy of Sciences. 2021;**118**(2)

[18] Wilcove DS, Rothstein D, Dubow J, Phillips A, Losos E. Quantifying threats

to imperiled species in the United States. Bioscience. 1998;**48**:607-615

[19] Zabel F, Delzeit R, Schneider JM, Seppelt R, Mauser W, Václavík T. Global impacts of future cropland expansion and intensification on agricultural markets and biodiversity. Nature Communications. 2019;**10**(1):1-10

[20] Raven PH, Wagner DL. Agricultural intensification and climate change are rapidly decreasing insect biodiversity. Proceedings of the National Academy of Sciences. 2021;**118**(2)

[21] Newbold T, Hudson LN, Hill SL, Contu S, Lysenko I, Senior RA, et al. Global effects of land use on local terrestrial biodiversity. Nature. 2015;**520**(7545):45-50

[22] Dudley N, Alexander S. Agriculture and biodiversity: A review. Biodiversity. 2017;**18**:45-49

[23] Stokstad E. New global study reveals the 'staggering' loss of forests caused by industrial agriculture. Science. 2018. Available from: https://www.sciencemag.org/news/2018/09/scientists-reveal-how-much-world-s-forests-being-destroyed-industrial-agriculture [Accessed date: 7 February 2020]

[24] Syktus JI, McAlpine CA. More than carbon sequestration: Biophysical climate benefits of restored savanna woodlands. Scientific Reports. 2016;**6**:29194

[25] Janzen DH, Hallwachs. Perspective: Where might be many tropical insects? Biological Conservation. 2019;**233**: 102-108

[26] Seibold S et al. Arthropod decline in grasslands and forests is associated with landscape-level drivers. Nature. 2019; **574**:671-674

[27] Wagner DL. Insect declines in the Anthropocene. Annual Review of Entomology. 2020;**65**:457-480

[28] Biesmeijer JC, Roberts SP, Reemer M, Ohlemüller R, Edwards M, Peeters T, et al. Parallel declines in pollinators and insect-pollinated plants in Britain and the Netherlands. Science. 2006;**313**(5785):351-354

[29] Powney GD, Carvell C, Edwards M, Morris RK, Roy HE, Woodcock BA, et al. Widespread losses of pollinating insects in Britain. Nature Communications. 2019;**10**(1):1-6

[30] Fox R, Asher J, Brereton T, Roy D, Warren MT. State of Butterflies in Britain and Ireland. Newbury, U.K.: Pisces Publications; 2006

[31] Erhardt A, Thomas JA. Lepidoptera as indicators of change in semi-natural grasslands of lowland and upland in Europe. In: Collins NM, Thomas J, editors. The Conservation of Insects and Their Habitats. London: Academic Press; 1991. p. 2130236

[32] Fox R. The decline of moths in Great Britain: a review of possible causes. Insect Conservation and Diversity. 2013;**6**:5-19

[33] Hahn M, Schotthöfer A, Schmitz J, Franke LA, Brühl CA. The effects of agrochemicals on Lepidoptera, with a focus on moths, and their pollination service in field margin habitats. Agriculture, Ecosystems and Environment. 2015;**207**:153-162

[34] Maes D, Van Dyck H. Butterfly diversity loss in Flanders (north Belgium): Europe's worst case scenario? Biological Conservation. 2001;**99**:263-276

[35] van Dyck H, van Strien AJ, Maes D, van Swaay CAM. Declines in common, widespread butterflies in a landscape under intense human use. Conservation Biology. 2009;**23**:957-965

[36] Groenendijk D, van der Meulen J. Conservation of moths in The

Netherlands: population trends, distribution patterns and monitoring techniques of day-flying moths. Journal of Insect Conservation. 2004;**8**:109-118

[37] Franzén M, Johannesson M. Predicting extinction risk of butterflies and moths (Macrolepidoptera) from distribution patterns and species characteristics. Journal of Insect Conservation. 2007;**11**:367-390

[38] Kuussaari M, Heliölä J, Pöyry J, Saarinen K. Contrasting trends of butterfly species preferring semi-natural grasslands, field margins and forest edges in northern Europe. Journal of Insect Conservation. 2007;**11**:351-366

[39] Mattila N, Kaitala V, Komonen A, Kotiaho Janne S, PÄIvinen, J. Ecological determinants of distribution decline and risk of extinction in moths. Conservation Biology. 2006;**20**: 1161-1168

[40] Melero Y, Stefanescu C, Pino J. General declines in Mediterranean butterflies over the last two decades are modulated by species traits. Biological Conservation. 2016;**201**:336-342

[41] van Swaay C, Warren M, Loïs G. Biotope use and trends of European butterflies. Journal of Insect Conservation. 2006;**10**:189-209

[42] van Swaay C, Cuttelod A, Collins S, Maes D, Munguira MLP, ŠaŠić M, et al. European Red List of Butterflies. Luxembourg: Publications Office of the European Union; 2010

[43] Swengel SR, Swengel AB. Assessing abundance patterns of specialized bog butterflies over 12 years in northern Wisconsin USA. Journal of Insect Conservation. 2015;**19**:293-304

[44] Forister ML, Cousens B, Harrison JG, Anderson K, Thorne JH, Waetjen D, et al. 2016

[45] Breed GA, Stichter S, Crone EE. Climate-driven changes in north-eastern US butterfly communities. Nature Climate Change. 2012;**3**:142

[46] Nakamura Y. Conservation of butterflies in Japan: Status, actions and strategy. Journal of Insect Conservation. 2011;**1**:5-22

[47] Chen IC, Hill JK, Shiu HJ, Holloway JD, Benedick S, Chey VK, et al. Asymmetric boundary shifts of tropical montane Lepidoptera over four decades of climate warming. Global Ecology and Biogeography. 2011;**20**: 34-45

[48] Gallai N, Salles JM, Settele J, Vaissière BE. Economic valuation of the vulnerability of world agriculture confronted with pollinator decline. Ecological Economics. 2009;**68**(3): 810-821

[49] Williams PH. The distribution and decline of British bumble bees (Bombus Latr.). Journal of Apicultural Research. 1982;**21**:236-245

[50] Goulson D, Hanley ME, Darvill B, Ellis JS, Knight ME. Causes of rarity in bumblebees. Biological Conservation. 2005;**122**:1-8

[51] Dupont YL, Damgaard C, Simonsen V. Quantitative historical change in bumblebee (Bombus spp.) assemblages of red clover fields. PLoS One. 2011;**6**:e25172

[52] Kosior A, Celary W, Olejniczak P, Fijal J, Król W, Solarz W, et al. The decline of the bumble bees and cuckoo bees (Hymenoptera: Apidae: Bombini) of Western and Central Europe. Oryx. 2007;**41**:79-88

[53] Bommarco R, Lundin O, Smith HG, Rundlöf M. Drastic historic shifts in bumble-bee community composition in Sweden. Proceedings of the Royal Society B: Biological Sciences. 2012;**279**:309-315

[54] Bommarco R, Kleijn D, Potts SG. Ecological intensification: harnessing ecosystem services for food security. Trends in Ecology & Evolution. 2013;**28**:230-238

[55] Colla S, Packer L. Evidence for decline in eastern North American bumblebees (Hymenoptera: Apidae), with special focus on Bombus affinis Cresson. Biodiversity and Conservation. 2008;**17**:1379-1391

[56] Thorp RW, Shepherd MD. Profile: Subgenus Bombus. In: Shepherd MD, Vaughan DM, Black SH, editors. Red List of Pollinator Insects of North America. Portland, Oregon: The Xerces Society for Invertebrate Conservation; 2005

[57] Cameron SA, Lozier JD, Strange JP, Koch JB, Cordes N, Solter LF, et al. Patterns of widespread decline in North American bumble bees. Proceedings of the National Academy of Sciences of the United States of America. 2011;**108**:662-667

[58] Ellis J. The honey bee crisis. Outlooks on Pest Management. 2012;**23**:35-40

[59] Ellis JD, Evans JD, Pettis J. Colony losses, managed colony population decline, and Colony Collapse Disorder in the United States. Journal of Apicultural Research. 2010;**49**:134-136

[60] Anderson KE, Sheehan TH, Eckholm BJ, Mott BM, DeGrandi-Hoffman G. An emerging paradigm of colony health: Microbial balance of the honey bee and hive (Apis mellifera). Insectes Sociaux. 2011;**58**:431-444

[61] Johnson RM, Dahlgren L, Siegfried BD, Ellis MD. Acaricide, fungicide and drug interactions in honey bees (Apis mellifera). PLoS One. 2013;**8**:e54092

[62] Huang Z. Pollen nutrition affects honey bee stress resistance. Terrestrial Arthropod Reviews. 2012;**5**:175-189

[63] Cooling M, Hoffmann BD. Here today, gone tomorrow: Declines and local extinctions of invasive ant populations in the absence of intervention. Biological Invasions. 2015;**17**:3351-3357

[64] Vogel V, Pedersen JS, Giraud T, Krieger MJB, Keller L. The worldwide expansion of the Argentine ant. Diversity and Distributions. 2010;**16**:170-186

[65] Wilson EO. 2002. The Future of Life. Abacus, Time Warner Book Group, London, UK. Sorvari, J., Hakkarainen, H. Wood ants are wood ants: deforestation causes population declines in the polydomous wood ant Formica aquilonia. Ecological Entomology. 2007;**32**:707-711

[66] Alburaki M, Chen D, Skinner JA, Meikle WG, Tarpy DR, Adamczyk J, et al. Honey bee survival and pathogen prevalence: From the perspective of landscape and exposure to pesticides. Insects. 2018:65

[67] Brandt A, Hohnheiser B, Sgolastra F, Bosch J, Meixner MD, Büchler R. Immunosuppression response to the neonicotinoid insecticide thiacloprid in females and males of the red mason bee Osmia bicornis L. Scientific Reports 2020;10(1):1-0.

[68] Williams GR, Troxler A, Retschnig G, Roth K, Yañez O, Shutler D, et al. Neonicotinoid pesticides severely affect honey bee queens. Scientific Reports. 2015;**5**(1):1-8

[69] Kairo G, Biron DG, Abdelkader FB, Bonnet M, Tchamitchian S, Cousin M, et al. Nosema ceranae, Fipronil and their combination compromise honey bee reproduction via changes in male

physiology. Scientific Reports. 2017;**7**(1):1-4

[70] Pettis JS, Rice N, Joselow K, vanEngelsdorp D, Chaimanee V. Colony failure linked to low sperm viability in honey bee (Apis mellifera) queens and an exploration of potential causative factors. PLoS One. 2016;**11**:e0147220

[71] Losey JE, Vaughan M. The economic value of ecological services provided by insects. Bioscience. 2006;**56**:311-323

[72] Sorvari J, Hakkarainen H. Wood ants are wood ants: deforestation causes population declines in the polydomous wood ant Formica aquilonia. Ecological Entomology. 2007;**32**(6):707-711

[73] Petanidou T, Vujic A, Ellis WN. Hoverfly diversity (Diptera: Syrphidae) in a Mediterranean scrub community near Athens, Greece. Annales de la Société entomologique de France. 2011;**47**:168-175

[74] Pearson DL, Cassola F. World-wide species richness patterns of tiger beetles (Coleoptera: Cicindelidae): Indicator taxon for biodiversity and conservation studies. Conservation Biology. 1992;**6**:376-391

[75] Desender K, Turin H. Loss of habitats and changes in the composition of the ground and tiger beetle fauna in four West European countries since 1950 (Coleoptera: Carabidae, Cicindelidae). Biological Conservation. 1989;**48**:277-294

[76] Turin H, Den Boer PJ. Changes in the distribution of carabid beetles in The Netherlands since II. Isolation of habitats and long-term time trends in the occurence of carabid species with different powers of dispersal (Coleoptera, Carabidae). Biological Conservation. 1988;**44**(3):179-200

[77] Brooks DR, Bater JE, Clark SJ, Monteith DT, Andrews C, Corbett SJ, et al. Large carabid beetle declines in a United Kingdom monitoring network increases evidence for a widespread loss in insect biodiversity. Journal of Applied Ecology. 2012;**49**(5):1009-1019

[78] Roulston TAH, Goodell K. The role of resources and risks in regulating wild bee populations. Annual Review of Entomology. 2011;**56**:293-312

[79] Biesmeijer JC, Roberts SPM, Reemer M, Ohlemuller R, Edwards M, Peeters T, et al. Parallel declines in pollinators and insect-pollinated plants in Britain and the Netherlands. Science. 2006;**313**:351-354

[80] Marlin JC, LaBerge WE. The native bee fauna of Carlinville, Illinois, revisited after 75 years: A case for persistence. Conservation Ecology. 2001;**5**

[81] Gardner JD, Spivak M. A survey and historical comparison of the Megachilidae (Hymenoptera: Apoidea) of Itasca State Park, Minnesota. Annals of the Entomological Society of America. 2014;**107**:983-993

[82] Bennett AB, Isaacs R. Landscape composition influences pollinators and pollination services in perennial biofuel plantings. Agriculture, Ecosystems & Environment. 2014;**193**:1-8

[83] Nemesio A. Are orchid bees at risk? First comparative survey suggests declining populations of forest-dependent species. Brazilian Journal of Biology. 2013;**73**:367-374

[84] Frankie GW, Rizzardi M, Vinson SB, Griswold TL. Decline in bee diversity and abundance from 1972-2004 on a flowering leguminous tree, Andira inermis in Costa Rica at the interface of disturbed dry forest and the urban environment. Journal of the Kansas Entomological Society. 2009;**82**(1):1-20

[85] Paukkunen J, Poyry J, Kuussaari M. Species traits explain long-term population trends of Finnish cuckoo wasps (Hymenoptera: Chrysididae). Insect Conservation and Diversity. 2018;**11**:58-71

[86] Sato S, Dixon AF. Effect of intraguild predation on the survival and development of three species of aphidophagous ladybirds: consequences for invasive species. Agricultural and Forest Entomology. 2004;**1**:21-24

[87] Brown M, Miller S. Coccinellidae (Coleoptera) in apple orchards of eastern West Virginia and the impact of invasion by Harmonia axyridis. Entomological News. 1998;**109**:143-151

[88] Stefanescu C, Aguado LO, Asís JD, Baños-Picón L, Cerdá X, García MAM, et al. Diversidad de insectos polinizadores en la peninsula ibérica. Ecosistemas: Revista Cietifica y Tecnica de Ecologia y Medio Ambiente. 2018;**27**:9-22

[89] Painter MK, Tennessen KJ, Richardson TD. Effects of repeated applications of Bacillus thuringiensis israelensis on the mosquito predator Erythemis simplicicollis (Odonata: Libellulidae) from hatching to final instar. Environmental Entomology. 1996;**25**(1):184-191

[90] Samways MJ. Diversity and conservation status of South African dragonflies (Odonata). Odonatologica. 1999;**28**(1):13-62

[91] Allan JD, Flecker AS. Biodiversity conservation in running waters. Bioscience. 1993;**43**:32-43

[92] Tierno de Figueroa JM, López-Rodríguez MJ, Lorenz A, Graf W, Schmidt-Kloiber A, Hering D. Vulnerable taxa of European Plecoptera (Insecta) in the context of climate change. Biodiversity and Conservation. 2010;**19**:1269-1277

[93] McCafferty PW, Lenat DR, Jacobus LM, Meyer MD. The mayflies (Ephemeroptera) of the Southeastern United States. Transactions of the American Entomological Society. 2010;**136**:221-233 (1890-)

[94] Jinguji H, Thuyet D, Ueda T, Watanabe H. Effect of imidacloprid and fipronil pesticide application on Sympetrum infuscatum (Libellulidae: Odonata) larvae and adults. Paddy and Water Environment. 2013;**11**:277-284

[95] Houghton DC, Holzenthal RW. Historical and contemporary biological diversity of Minnesota caddisflies: A case study of landscape-level species loss and trophic composition shift. Journal of the North American Benthological Society. 2010;**29**:480-495

[96] McGuinness CA. Carabid beetle (Coleoptera: Carabidae) conservation in New Zealand. Journal of Insect Conservation. 2007;**11**:31-41

[97] Harmon JP, Stephens E, Losey J. The decline of native coccinellids (Coleoptera: Coccinellidae) in the United States and Canada. Journal of Insect Conservation. 2007;**11**:85-94

[98] Wheeler AG, Hoebeke ER. Rise and fall of an immigrant lady beetle: Is Coccinella undecimpunctata L.(Coleoptera: Coccinellidae) still present in North America? Proceedings of the Entomological Society of Washington. 2008;**110**(3):817-823

[99] Brown PM, Roy HE. Native ladybird decline caused by the invasive harlequin ladybird Harmonia axyridis: Evidence from a long-term field study. Insect Conservation and Diversity. 2018;**3**:230-239

[100] Camacho-Cervantes M, Ortega-Iturriaga A, Del-Val E. From effective biocontrol agent to successful invader: the harlequin ladybird (Harmonia axyridis) as an example of

good ideas that could go wrong. PeerJ. 2017;**5**:e3296

[101] Rutledge CE, O'Neil RJ, Fox TB, Landis DA. Soybean aphid predators and their use in Integrated Pest Management. Annals of the Entomological Society of America. 2004;**97**:240-248

[102] Honek A, Martinkova Z, Kindlmann P, Ameixa Olga MCC, Dixon Anthony FG. Long-term trends in the composition of aphidophagous coccinellid communities in Central Europe. Insect Conservation and Diversity. 2014;**7**:55-63

[103] Lobo JM. Decline of roller dung beetle (Scarabaeinae) populations in the Iberian peninsula during the 20th century. Biological Conservation. 2001;**97**:43-50

[104] Carpaneto GM, Mazziotta A, Valerio L. Inferring species decline from collection records: roller dung beetles in Italy (Coleoptera, Scarabaeidae). Diversity and Distributions. 2007;**13**:903-919

[105] Lumaret JP, Galante E, Lumbreras C, Mena J, Bertrand M, Bernal JL, et al. Field effects of ivermectin residues on dung beetles. Journal of Applied Ecology. 1993;**30**:428-436

[106] Relyea RA, Hoverman JT. Interactive effects of predators and a pesticide on aquatic communities. Oikos. 2008;**117**:1647-1658

[107] Lumaret J-P. Atlas des Coléoptères Scara-béides Laparosticti de France. Paris, France: Secrétariat Faune Flore/ MNHN; 1990

[108] Lobo JM, Lumaret J-P, Jay-Robert P. Diversity, distinctiveness and conservation status of the Mediterranean coastal dung beetle assemblage in the Regional Natural Park of the Camargue (France). Diversity and Distributions. 2001;**7**:257-270

[109] Nieto A, Alexander KN. The Status and Conservation of Saproxylic Beetles in Europe. 2010

[110] Lindhe A, Jeppsson T, Ehnstrom B. Longhorn beetles in Sweden - changes in distribution and abundance over the last two hundred years. Entomologisk Tidskrift. 2011;**131**:507

[111] Schuch S, Wesche K, Schaefer M. Long-term decline in the abundance of leafhoppers and planthoppers (Auchenorrhyncha) in Central European protected dry grasslands. Biological Conservation. 2012;**149**:75-83

[112] Schuch S, Bock J, Leuschner C, Schaefer M, Wesche K. Minor changes in orthopteran assemblages of Central European protected dry grasslands during the last 40 years. Journal of Insect Conservation. 2011;**15**:811-822

[113] Relyea RA, Hoverman JT. Interactive effects of predators and a pesticide on aquatic communities. Oikos. 2008;**117**(11):1647-1658

[114] Kalkman VJ, Boudot J-P, Bernard R, Conze K-JR, Knijf GD, Dyatlova E, et al. European Red List of Dragonflies. Luxembourg: Publications Office of the European Union; 2010

[115] DeWalt RE, Favret C, Webb DW. Just how imperiled are aquatic insects? A case study of stoneflies (Plecoptera) in Illinois. Annals of the Entomological Society of America. 2005;**98**:941-950

[116] Clausnitzer V, Kalkman VJ, Ram M, Collen B, Baillie JEM, Bedjanič M, et al. Odonata enter the biodiversity crisis debate: The first global assessment of an insect group. Biological Conservation. 2009;**142**:1864-1869

[117] Ball-Damerow JE, M'Gonigle LK, Resh VH. Changes in occurrence,

richness, and biological traits of dragonflies and damselflies (Odonata) in California and Nevada over the past century. Biodiversity and Conservation. 2014;**23**:2107-2126

[118] Kadoya T, Suda SI, Washitani I. Dragonfly crisis in Japan: A likely consequence of recent agricultural habitat degradation. Biological Conservation. 2009;**142**(9):1899-1905

[119] Futahashi R. Diversity of UV reflection patterns in Odonata. Frontiers in Ecology and Evolution. 2020;**8**:201

[120] Nakanishi K, Nishida T, Kon M, Sawada H. Effects of environmental factors on the species composition of aquatic insects in irrigation ponds. Entomological Science. 2014;**17**:251-261

[121] Zwick P. Stream habitat fragmentation — a threat to biodiversity. Biodiversity and Conservation. 1992;**1**:80-97

[122] Zwick P. Phylogenetic system and zoogeography of the Plecoptera. Annual Review of Entomology. 2000;**45**:709-746

[123] Fochetti R, de Figueroa JMT. Notes on diversity and conservation of the European fauna of Plecoptera (Insecta). Journal of Natural History. 2006;**40**:2361-2369

[124] Tixier G, Guérold F. Plecoptera response to acidification in several headwater streams in the Vosges Mountains (northeastern France). Biodiversity and Conservation. 2005;**14**:1525-1539

[125] Bojková J, Komprdová K, Soldán T, Zahrádková S. Species loss of stoneflies (Plecoptera) in the Czech Republic during the 20th century. Freshwater Biology. 2012;**57**:2550-2567

[126] Bojková J, Rádková V, Soldán T, Zahrádková S. Trends in species diversity of lotic stoneflies (Plecoptera) in the Czech Republic over five decades. Insect Conservation and Diversity. 2014;**7**:252-262

[127] Nedbalová L, Vrba J, Fott J, Kohout L, Kopáček J, Macek M, et al. Biological recovery of the Bohemian Forest lakes from acidification. Biologia. 2006;**61**(20):S453-S465

[128] Kury D. Changes in the ephemeroptera and plecol_ftera population s of a swiss j ura stream (roserenbach) between 1935 and 1990

[129] Zahrádková S, Soldán T, Bojková J, Helešic J, Janovská H, Sroka P. Distribution and biology of mayflies (Ephemeroptera) of the Czech Republic: present status and perspectives. Aquatic Insects. 2009;**31**:629-652

[130] Zedková B, Rádková V, Bojková J, Soldán T, Zahrádková S. Mayflies (Ephemeroptera) as indicators of environmental changes in the past five decades: A case study from the Morava and Odra River Basins (Czech Republic). Aquatic Conservation. 2015;**25**:622-638

[131] Pescador ML, Lenat DR, Hubbard MD. Mayflies (Ephemeroptera) of North Carolina and South Carolina: an update. Florida Entomologist. 1999:316-332

[132] McCafferty PW. Status of some historically unfamiliar American mayflies (Ephemeroptera). Pan-Pacific Entomologist. 2001;**77**:210-218

[133] Jenderedjian K, Hakobyan S, Stapanian MA. Trends in benthic macroinvertebrate community biomass and energy budgets in Lake Sevan, 1928-2004. Environmental Monitoring and Assessment. 2012;**184**:6647-6671

[134] Karatayev AY, Burlakova LE, Padilla DK, Mastitsky SE, Olenin S. Invaders are not a random selection of species. Biological Invasions. 2009;**11**(9)

[135] Chamberlain DE, Fuller RJ, Bunce RG, Duckworth JC, Shrubb M. Changes in the abundance of farmland birds in relation to the timing of agricultural intensification in England and Wales. Journal of Applied Ecology. 2000;**37**(5):771-788

[136] Jabbar A, Wu Q, Peng J, Zhang J, Imran A, Yao L. Synergies and determinants of sustainable intensification practices in Pakistan agriculture. Landscape. 2020;**9**:10. DOI: 10:33390/Land9040110

[137] Haile B, Cox C, Azzarri C, Koo J. Adoption of sustainable intensification practices: Evidence from maize-legume farming system in Tanzania. In: IFFPRI Discussion paer 01696. International Food Policy Research Institute; 2017

[138] Pretty J, Bharucha ZP. Sustainable intensification in agricultural systems. Annals of Botany. 2014;**114**:1571-1596. DOI: 10.1093/aob/mcu205

[139] Wyckhuys KA, Hughes AC, Buamas C, Johnson AC, Vasseur L, Reymondin L, et al. Biological control of an agricultural pest protects tropical forests. Communications Biology. 2019;**2**(1):1-8

[140] van Strien AJ, Meyling AW, Herder JE, Hollander H, Kalkman VJ, Poot MJ, et al. Modest recovery of biodiversity in a western European country: The living planet index for the Netherlands. Biological Conservation. 2016;**200**:44-50

[141] Arzate S, Sánchez JG, Soriano-Molina P, López JC, Campos-Mañas MC, Agüera A, et al. Effect of residence time on micropollutant removal in WWTP secondary effluents by continuous solar photo-Fenton process in raceway pond reactors. Chemical Engineering Journal. 2017;**316**:1114-1121

[142] Pascal-Lorber S, Laurent F. Phytoremediation techniques for pesticide contaminations. In: Alternative Farming Systems, Biotechnology, Drought Stress and Ecological Fertilisation. Dordrecht: Springer; 2011. pp. 77-105

Chapter 4

# Pesticide Impact on Honeybees Declines and Emerging Food Security Crisis

*Farkhanda Manzoor and Mahnoor Pervez*

## Abstract

Bee crisis is threatening worldwide food security. Pesticides are extensively used in the agricultural zone. Unfortunately, these pesticides cause severe toxicity toward pollinators than the target pests such as honeybees. This review summarizes the different studies related to pesticide hazards of bees. This paper reported risks of pesticides neurological and physiological poisoning toward honeybees. Pesticides act as poison and ruin vital functions involved in leaning and cognition, behavior and, the body physiological mechanisms. Many laboratory and field research data evaluated the lethal and sub-lethal poisoning on bee foraging dance, learning, and memory abilities of honeybees. Insecticide residues are detected in bee bodies and $LD_{50}$ and $LC_{50}$ values evaluated. It is also studied that in honeybees systemic insecticide residues and, its metabolite adulterated in their body during foraging activities. Similarly, pesticide-contaminated food stored in a hive consumed continuously by honeybees may cause sub-lethal toxicity effects. Which causes anomalous bee social behavior and ultimately leads to colony collapse disorder. If population of pollinator decline it will disturb the food chain and leads to food crisis. This review emphasized causes of bee decline with the emergence of pesticides in agricultural domains.

**Keywords:** Honeybee (*A. mellifera*), Pesticide poisoning, Pollinator decline, Route of exposure, Lethal effects, Sub-lethal effects

## 1. Introduction

Honeybees are the most important and economically dominant pollinator species for food crop production over the global [1]. It has been studied that 87.5% of flowering and edible plants are pollinated by animals such as honeybees. Honeybees also produce wax, honey, and venom. Economic global estimate concluded that a global economic value of €88bn was associated with bees pollinating crops and products [2]. Afterward, this is revised and estimate and recorded an increase of the economic value of €115bn [3]. This raising trend highlights the dependence upon pollinators in the global food supply. However, researchers noticed the disappearance of bees and their population has been decline [4]. Most important crops, fruits, vegetables, and fodder crops are badly affected by the decline in the honeybee population. Researchers have great concern about the honeybee population decline [5, 6]. In the United States 30–40% disappearance of the honeybee colonies attributed to colony collapse disorder A disease syndrome described as the sudden and speedy loss of honeybees population [7]. In the United Kingdom, 54% honeybee population lost in the last decades [8]. China has 6 million bee colonies of honeybees. Chinese beekeepers have faced inexplicable colony losses and a decline in the bee population [9]. Various factors such as biotic (parasites and pathogens) and abiotic (climate changes, habitat destruction) are blamed for this bee decline [10]. *Varroa* mites are a serious threat to apiculture globally. *Varroa* mites feed on bee hemolymph and are linked to winter loss of bee colonies. In Germany, it was found that *Varroa* infestation, viral and bacterial infections were associated with winter loss of honeybee colonies [11]. Similarly microsporidium also a worldwide honeybee pathogen [12]. Climatic factors such as increasing temperature, variation in rainfall patterns, and other extreme weather conditions have drastic effects on bees. Changing climatic conditions affect the interaction between pollinators and plants by changing the period of flower blooming [13]. Among various agents, pesticides are the most obvious and significant agent toward the loss of honeybees. They are directly responsible for bee poisoning and bee death [14]. This paper focusing on the (1) pesticide toxicity toward honeybees and route of exposure (2) occurrence of insecticide residues in pollen grains, bee bodies, hive wax, and honey (3) lethal and sub-lethal insecticide toxicity on honeybees. Various laboratory studies described the lethal and sub-lethal pesticide toxicity on honeybee learning, memory, food foraging and, physiological function. While few studies were observed in field studies. This chapter will emphasize the causes of bee decline with the emergence of pesticides in agricultural domains.

## 2. Plants preference for honeybees

Several plant species depend on insect pollinators for seed set honeybees consume pollen and nectar of sunflower, oilseed rape, maize, etc. These plants can possess threatened honeybees if they are treated with systematic pesticides. Bee mortality is observed when bees contact with pesticide treated plants and consume flower pollen and sap plant or nectar e.g. sunflower, *Helianthus annuus* L. [15, 16], oilseed rape, *Brassica napus* L. [17], and *Phacelia tanacetifolia* Benth. [18, 19] which bees consume [20]. It is also observed that pesticide effects on pollinator reproductive success and plant seed set potential. In Canada forest plants were treated with Matacil (aminocarb insecticide) to control spruce budworm. After pesticide application, foraging honeybees reduce, and high mortality was observed. Shortly, afterward as honeybee population reduce a large number of native plant diversity adversely reduce such as lilies flowers showed a reduction in fecundity. Commercial blueberry fields also suffered reduce fruit sets when adjacent lands were sprayed with pesticides.

## 3. Multiple routes of pesticide exposure

An ultimate issue of pesticide poisoning to honeybees is the assessment of the exposure scenario. Honeybees are exposed to insecticides through the following routes.

### 3.1 Exposure via direct contact

Honeybees form direct contact with pesticides. The most common way of pesticide exposure is the aerial spray. They are straightly contaminated while foraging during spray treatment in a field. Treatment also influences by wind drift. Pesticide droplets made direct contact with the honeybee body in the air. In 2003 it was observed that the bee population decline due to direct contact of dust emitted from pneumatic drilling machine used to plant neonicotinoid coated seeds [21]. Bees flying over sowing fields directly exposed to pesticides at lethal levels especially under humid conditions. In Humid environments pesticides, dust particles stick to the bee abdomen [22]. Herbicides and fungicides were directly applied to the soil before sowing of crops and most insecticides were applied on crops through aerial spraying. Pesticide aerial droplets and dust pesticide particles can make direct contact with flying honeybees and create toxicity. Fine droplets may also carry hundreds of meters from treated site to untreated sites with wind, and cause bee mortality. In addition to food, bees make direct contact with pesticides during water drinking. Pesticide residues from nearby treated fields leach into the water resources or mix with water by drifting from spray applications. Honeybees drink water from trenches, water pools, and rivers, if these water resources are polluted with pesticide, the foraging bees will ingest them.

### 3.2 Exposure via indirect contact or ingestion of pesticide residue

A systematic characteristic of pesticides provides the translocation of pesticide residues inside all plant tissues, phloem, and nectar. Consequently, honeybees are likely to be vulnerable to the pesticide while feeding on pollen grains and flower nectar [23]. The pesticide used for seed dressing can also possess a noticeable threat to pollinators as they are transported through the plants and contaminate the pollen [17]. In the same way, Early morning when plants produced guttation drops and systematic pesticides appear in such guttation drops in a dose that are lethal to bees [23]. Several studies showed the existence of insecticide residues in seeds, pollen, and nectars. The pesticide used for the treatment of seeds can be transported throughout growing plants and contaminate the nectars and pollens. The presence of imidacloprid residues (3 μg/kg) in pollen grains has been reported in Gaucho seed dresses sunflower [17]. Similarly, in France, a survey study of pesticide residues in pollen loads has been conducted. A survey report was showing the presence of imidacloprid and its metabolites nicotic acid (49% of 81 analyzed samples), Fipronil, and its metabolites fipronil sulfone and fipronil disulfinile (12% of 81 analyzed samples) [24]. Freshly stored pollen and bee bread are considered to be the principal source of in-hive contamination for adult and larvae bees [25]. In the same way, miticides and fungicide residues are well reported in pollen grain [26]. In Germany, thiacloprid was the extensively abundant pesticide as study results show that it was present in about 33% of the pollen samples at concentration levels up to 199 μg/kg [12]. It is investigated the level of pesticide in the droplets exude by the treated plants [27]. Three different systematic pesticides (imidacloprid, clothianidin, and thiamethoxam) and on the non-systematic pesticide (fipronil) were tested in field and lab trials. Analysis of drops showed that systematic pesticide concentration near to that applied in field sprays. Systematic pesticides found in pollen samples and guttation drops compared to nonsystematic pesticides. In apiaries across Spain, both acaricides and agricultural pesticides were found in beebread. Cypermethrin, deltamethrin, and chlorpyrifos were found at sub-lethal levels [28]. In Slovenia, honeybee colonies present near insecticides treated apple orchards showed insecticide residues in beebread [24]. Pollen samples were collected from different areas of North America. About 5.4% of pollen samples were contaminated with thiacloprid and acetamiprid. Similarly, about 1.9% of bee wax samples were contaminated with thiacloprid [29].

## 4. Chronic and acute toxicity

Chemical pesticide hazards are stated as acute and chronic toxicity. In acute poisoning, the toxic reactions are more violent and occur unexpectedly. Honeybees

| Pesticide | Exposure route | Toxicity $LD_{50}$ | References |
|---|---|---|---|
| Imidacloprid | Direct contact | 18 to 104 ng/bee | [36] |
| Imidacloprid | Ingestion | 4 to 60 ng/bee | [19, 37] |
| Imidacloprid | oral routes | 0.0037 μg/bee | [38] |
| Imidacloprid | topical application | 0.081 μg/bee | [37] |
| Clothianidin | Oral route | 21.8 ng/bee | [34] |
| Dinotefuran | Oral route | 75.0 ng/bee | [34] |
| Nitenpryan | Oral route | 13.8 ng/bee | [34] |
| Fipronil | Oral route | 3.45–3.86 ng/bee | [39] |
| Thiacloprid | Contact and Oral route | 24.2 μg/bee | [40] |
| Acetamprid | Contact route | 7.07 μg/bee | [34] |
| Oleofin | Oral route | 50 μg/bee | [16] |
| Thiamethoxam | Oral route | 0.004 μg/bee | [38] |

**Table 1.**
*Reported lethal effect toxicity of different pesticides toward honeybee workers.*

usually die from contact and ingestion of pesticides [30]. While in case of chronic exposure pesticide absorbed in small and repeated doses, the reaction usually slow but continuous, they may lead to the elimination of the entire colony [31, 32]. Insecticide toxicity generally measured using $LD_{50}$ values (Exposure concentration that create death of 50% of the exposed treated population [33]. Pesticide chemical toxicity thresholds for honeybees are generally set as high toxicity (acute $LD_{50}$ < 2 μg/honeybee), Moderate toxicity (acute $LD_{50}$: 2–10.99 μg/honeybee), Slightly toxic (acute $LD_{50}$: 11–100 μg/honeybee), Non-toxic (acute $LD_{50}$ > 100 μg/honeybee). According to the European regulation on risk valuation of pesticides on honeybees, the toxicity is evaluated by acute $LD_{50}$ and $LC_{50}$ values. These two parameters measure pesticide poisoning by calculating the number of dead bees that die after 24 hours of treatment. Therefore, the research and knowledge of pesticide toxicity on non-target species is vital to eliminate various agricultural pests without hurting the bees [34, 35]. **Table 1** shows reported $LD_{50}$ values of various pesticides toward honeybees.

## 5. Lethal and sub-lethal pesticide toxicity on honeybees

### 5.1 Lethal effects

Pesticide lethal toxicity is the major cause of honeybee decline. Detection of pesticide residues inside the honeybee body is a measure of pesticide poisoning to estimate lethal effects. Pesticides possess high toxicity due to specific mode of action. Neonicotinoids show delay toxicity with sub-lethal effects at low concentrations [11] or they cause the killing of bees if exposed for a long interval of time [41]. Similarly, neonicotinoids and fipronil suppress the immune system and vulnerable them to various infections such as *Nosema* infection [42] and the outbreak of *Varroa* mites. The lethal effects of certain insecticides are enhanced in the presence of fungicides, which act as synergists. These synergistic mixtures inhibit the bee detoxification system. In another study pesticide residues inside honeybee bodies. About 11.2% of honeybee samples were contaminated with imidacloprid, while in about 18.7% samples the dominant metabolite 6-chloronicotinic acid was detected.

Pesticide residue analysis is done to determine bee poisoning [43]. In the UK fungicide and herbicide residues are found in bee bodies. A study carried out by using gas chromatography along with tandem mass spectrometry for the detection of insecticides impurities and its metabolites in honeybees. About more than 200 insecticide and insecticide metabolites were found in tested honeybee body. Clothianidin ranging from 0.5 ng/g- 1.0 ng/g was detected in poisoned honeybees. This indicates the cause of honeybee death, similarly imidacloprid and thiamethoxam residues found in poisoned honeybee [44]. The detected imidacloprid concentration ranging from 0.3 ng/g-240.6 ng/g while thiamethoxam was detected in concentration ranging from 0.5-275 ng/ [45]. Similarly, a second abundant pesticide is a chlorpyrifos in tested honeybee samples. 1.5 ng/g concentration detected in tested honeybee bodies [46]. In Poland during early autumn there was a considerable decline in the honeybee population, it was not having any historical with spray application. An investigation confirmed that these toxic hives were located near cauliflower and brassica fields. The EU allowed seed treatment with sowing. These pesticide residues present in the guttation drops and nectar enter honeybee bodies during foraging activities [28].

### 5.2 Sub-lethal toxicity

Toxic effects of insecticides can be evaluated as sub-lethal toxicity when exposure to the insecticide does not induce death in the experimental population but may disturb biochemical, physiological, and behavioral changes such as by impairing foraging and learning behavior or disrupting another aspect of neurological and physiological functions [43]. Bees exposure at sub-lethal doses also have a negative impact on flower and scent recognition, the spatial orientation of bees, perturbations of the foraging patterns of the honeybee by disturbing navigation memory. In an experimental study, it was demonstrated contact of the honeybees to sub-lethal concentrations of thiamethoxam cause impairment of memory, brain and gut functions which ultimately lead to a shorter lifespan [47]. In laboratory analysis, sub-lethal concentration of deltamethrin disturbed body functions in honeybee at cellular level such as by causing marked dysfunctions in cardiac cells by changing frequency and muscle contraction [43]. The respiratory system is also badly affected after exposure to pesticides. Imidaloprid at sub-lethal exposure created marked changes in the respiratory pattern of bees and also in hypopharyngeal glands growing smaller in size as compared to untreated bees. Similarly, mobility behavior also affects after exposure to low doses of imidacloprid.

Negative impacts on mobility are dose-dependent and change with time [48, 49]. Honeybee foraging and spatial orientation is totally depend upon visual remembering or learning of landmarks. During foraging trips, bees memorized landmarks and used to navigate nectar and pollen sources as well as to communicate accurately to the rest of the colony members about direction of food source and its distance from hive. Exposure to pesticides disturbed the learning of visual landmarks during foraging round trips and the transfer of this information to other bees of the hive. The pyrethroid such as deltamethrin has been studied to change the foraging round trips of worker bees after treated topically with sub-lethal concentrations [50]. Bees exposed to the thiamethoxam treated pollen and nectar under semi-field conditions seem to reduce foraging trips and lost their ways. As a result, the colony becomes weaker and putting it at greater risk of collapse [31]. Similarly, fipronil applied topically at low doses to honeybees reduces their ability to sense sucrose smell by about 40% relative to the capacity of untreated bees [37]. Neonicotinoid pesticides have a toxic effects on the queen bee. They affect queen bee life span and egg-laying capacity. Thus colony size reduced and lead to colony collapse. Pesticides also repel

| Pesticide | Sub lethal toxicity | Reference |
|---|---|---|
| Deltamethrin | Cardiac dysfunctions in the heart cells | [43] |
| Imidacloprid | Shorter queen life span and reduced egg lying capacity and colony size, Disrupt development | [52] |
| Imidacloprid | Mobility and olfactory behavior | [51] |
| Deltamethrin | Homing trips disturbed | [53] |
| Thiamethoxam | Colony become weaker, affect foraging behavior, Impair brain and gut functions | [31, 50] |
| Fipronil | Reduce capacity of bees to detect food sources, olfactory learning and memory | [43, 54] |
| Diazon | Reduce reproduction and survival of queen, colony maintenance and foraging behavior. | [37] |

**Table 2.**
*Reported sub lethal toxicity of various pesticides toward honeybee.*

the pollinator from treated crops. If pollinators avoid foraging the pesticide contaminated crops and flowers this would adversely influence crop yields dependent upon the pollinator for pollination [51]. **Table 2** showed the sub-lethal toxicity of various pesticides toward honeybees.

## 6. Neurotoxic action of pesticides on bees

Pesticides react at their molecular target site, which may cause to effects impairing behavior and body physiology. Mostly pesticides act as a neurotoxin and disturb normal neurological function. Pesticides showed their lethal and sub-lethal poisoning by acting as an acetylcholinesterase inhibitor, nicotinic acetylcholine receptor agonists, voltage-gated $Na^+$ channel agonists, and GABA gated $Cl^-$ channel agonist.

### 6.1 GABA gated $Cl^-$ channel agonist

Fipronil is a phenyl pyrozal insecticide class. When bees are exposed to fipronil reached to target site in brain and binds to the GABA gated $Cl^-$ channels, it prevent the $Cl^-$ channels from closing and maintained it in open. This state leads to hyperpolarization and the inability to transmit action potential. Phenyl pyrozal is a systemic insecticide and is detected in pollen at 1-4 ppb concentration [53].

### 6.2 Nicotinic acetylcholine receptor agonists

Nicotinic mimics the action of acetylcholine, neurotransmitter, inside the body it react with nicotinic acetylcholine receptor (nAChR) and bind to receptor. After binding it stimulate the repeated generation of action. The new class of pesticides neonicotinoids is synthetic analogs of nicotine with a higher affinity for the nAChR in the bee brain [16]. A study showed that neonicotinoid pesticides such as imidacloprid, clothianidin, and thiamethoxam are highly toxic to bees with acute $LD_{50}$ from 0.004 to 0.075 μg/bees [55]. Imidacloprid metabolites such as 5-hydroxyimidacloprid and olefin have a high affinity for the honeybee nAChR [16]. Another neonicotinoid insecticide thiamethoxam also has affinity for nACh receptors, however, thiamethoxam rapidly degraded in to high affinity clothianidin metabolites. All neonicotinoid pesticides damage the capability of worker bees to forage and go back to the hive [56].

### 6.3 Acetylcholinesterase inhibitor

Organophosphate and methylcarbamate pesticides behave as neurotoxin and inhibit the AChE enzyme activity. AChE enzyme deactivates the acetylcholine neurotransmitter at neuron synapses [57]. Neuro pesticides have wide range of bee poisoning ($LD_{50}$ = 0.018–31.2 μg/bees) [58]. In United State, about 117 bee poisoning incidents were investigated between 1994 and 2003. The maximum number of these poisoning incidents attributed to dimethoate and bendicarb [59]. Similarly, coumaphos is an organophosphate that is used by beekeepers against varroa mites [60] When beekeepers repeatedly used coumaphos it is concentrated and build up in the colony wax [61]. Poisoning of bee colonies with coumaphos is linked with significant mortality of bee queens and worker bees. It studied that bee larvae raised on food contaminated with 8 mg/L coumaphos were showed more mortality rate than control larvae [62].

### 6.4 Voltage-gated $Na^+$ channel agonists

A pyrethroid is an extensively used synthetic insecticides. Pyrethroids and DDT are organophosphate. They are neurotoxin and their target site inside brain is voltage-gated $Na^+$ channels in the axon of the nerve cell. At their target site pyrethroids delay the closing of Na + channels and prolonged the recovery period after generation of the action potential. The pyrethroid flumethrin and tau-fluvalinate are extensively used to control *varroa* mites, it may be accumulated in bee wax as high as 200 ppm and cause bee death [63].

## 7. Behavior disturbs by pesticide exposure

### 7.1 Habituation

Habituation is a learning behavior that can be explain as steady reduction in the frequency of response to a repeated or useless stimulus. Habituation enable an individuals to tolerant the repeated stimulus and save energy. It is a predominantly vital practice for the honeybees because it permits them to escape both unusable and tedious stimuli. Thus, save time and energy. In honeybee habituation behavior involves the proboscis extension response (PER) stimulated by an antenna touch with a sugar solution. The frequency of antennal contacts with sugar solution calculated to perceive a detention of proboscis extension. Honeybee exposed to neonicotinoid pesticide at sub-lethal concentrations such as imidacloprid enhanced the PER habituation [51]. In another experiment, seven to eight days olds bees were used to elicit antennal receptor stimulations at shorter time intervals [64]. In this research trial approximately 7 and 8 days old bees were used, because in this age bees started their short orientation flight. At start of the exposure (start one hours), imidacloprid upsurges the number of trials required to attain habituation in 7-day-old bees but declines it in 8-day-old bees. As bees become aged or older the habituation behavior become more refine with time. The habituation profile obtained with imidacloprid evolves with time particularly. The development of the habituation learning with time due to the action of imidacloprid metabolites olefin and 5-hydroxyimidacloprid that pause and increase habituation, respectively [63]. These results proposed the presence of two different nicotinic acetylcholine receptors (nAChR), that are differentially expressed as young bees mature.

### 7.2 Disturb learning and memory

In social insects honeybee learning and memory is an important behavior for absolute adaptation of the individual to their environment. This behavior enables the honeybees to fulfill the colony requirement. Various pesticides have lethal and sub-lethal toxicity on the honeybee brain and interrupt honeybee learning and memory behavior. A sub-lethal dose of parathion induced an alteration in time of foraging to the early morning. This modification in foraging time can be elucidated by the alteration in the circadian clock [65]. In one study toxic effects of OP methyl-parathion on honeybee foraging, visual and olfactory tasks have been described [66]. Since methyl-parathion is recognized as an acetylcholinesterase inhibitor. The methyl-parathion toxicity stimulated actions has been understood in terms of the advance of cholinergic signaling by inhibiting AChE. Proboscis extension reflex (PER) responses have been studied to evaluate the honeybee learning and memory process. This allows the path to streamlined laboratory procedures and studied the various memory phases. Bees exposed to sub-lethal doses of pyrethroid through tarsal contact may show a weakening of the conditioned responses, which point out that during forging activities pesticide residual contact impair learning and cognitive process [67]. In another laboratory test, the honeybees were feed on different pyrethroid insecticides contaminated sucrose solution for 11 days. It was concluded that deltamethrin impairs the proboscis extension responses, whereas λ-cyhalothrin, cypermethrin, and τ-fluvalinate did not disturb behavior patterns. This demonstrated that lethal toxicity induced by pesticides is more substance-specific than family-specific. Similarly, under a semi-field environment, the presence of deltamethrin prompts a significant decrease in the honeybee foraging trips, which is retreated when the pesticide contact terminates. Results inferred that deltamethrin was considered to be lethal in this study. Similarly, in another study both lethal and sub-lethal doses of imidacloprid elicits a reduction in learning activities, studied by the conditioned PER responses [68]. This effect also depends on the seasonal pattern, as results showed that during the summer season honeybees are more sensitive to imidacloprid, compared to in winter. In the semi-field experiment, the imidacloprid residues in foraging plants generate a considerable decline in the foraging bees. The decline in the foraging bee population is linked with a falling-off of olfactory and learning behavior. T-tube maze is a behavioral approach that is associated with learning by integrating visual and spatial orientation. Honeybees that are exposed to sublethal doses of imidacloprid in a T-tube maze showed a considerable reduction in visual learning capacity and diminished olfactory responses in the PER assay [69]. The toxicity of fipronil on honeybee learning and memory behavior has also been explained by olfactory conditioning of the PER. Sub-chronic exposure of bees to fipronil causes a considerable decrease in the learning process [70]. Results showed that cuticle contact of a sub-lethal dose weakens the olfactory memory whereas a lethal dose administration through thorax changes antennal tactile learning [71].

### 7.3 Olfaction and gustation

Olfaction and gustation are vital senses in the life of the honeybee colony both at the individual and colony levels. Both responses are trigger and processed at the neural level [69]. They are intricate in odor and sense of taste recognition which is required for the bee to forage and visit flowers, identify foreign bees in the hive, forager workers, social communication for social cohesion of the colony and, recognize allelochemicals and nectar in plants [72, 73]. Pesticides may disturb the honeybee olfactory and gustatory senses by disturbing physiological processes control scent recognition or other neuronal signals [74, 75]. The toxicity of pesticides on gustation can be estimated by reviewing the gustatory threshold modules, which is defined as the minimum dose of a sugar solution make contact with antenna and able to elicit a PER. Bees topically treated with imidacloprid show an escalation of the gustatory threshold [51]. This effect is upsurges with time. This contrasts with acetamiprid that upturns sensitivity of the antennal stimulation sugar solution by oral route. Thiamethoxam induces no toxicity on sugar sensitivity at the lowest concentration [76] and a reduction in sucrose stimulation detected at higher concentrations [72].

### 7.4 Navigation and orientation

Navigation and orientation are important which enable the honeybees to collect the food vital for colony development. It is an integrative component of vision, olfaction, orientation and signal treatment [77, 78]. Navigation involves the food source location communication with other colony members, chiefly when foraging site located at a distance of more than 100 m. The influences of insecticides on the foraging communication dance have been studied in various experiments. Parathion avoids the honeybees to share the correct location of the foraging site by shifting the wrong angle during communication dance. Forager bees showed shorten distances through an intensification of waggle dance rhythm, with no alteration on horizontal and vertical combs. The foraging communication error may be due to a disturbance in neural function [54, 79]. The homing flight is the flight made by the bees from the foraging sites to the hive location. It is a return flight. This flight pattern is used in evaluating the adverse sub-lethal effects of pesticides on honeybees. Deltamethrin hampers return flight or homing activity behavior. Various approaches have been developed for studying honeybee movements and flight in diverse space scales and dimensions [80]. Harmonic radar and radiofrequency identification are two approaches that appear appropriate for investigating navigation and return flight to home. In harmonic radar approach bees are fitted with a transponder to analyze movement in the landscape. This approach allowing researchers to monitor free-flying bees [81]. This approach is utilized to detect specific behaviors of orientation and navigational flight, particularly trip duration related with flight speed that escalate with time [82] and to establish the use of map-like spatial memory in honeybee navigation [83]. Another approach is based on radiofrequency identification (RFID) [84]. Honeybees are tagged with a passive RFID microchip that releases a specific individualized radio signal after activation by a radio frequency. RFID technique widely applied to evaluate the foraging behavior and the circadian foraging rhythm of bees [85]. This technique has been shown that fipronil leads to a decline in the number of individual foraging flights and increased the homing flight time. In the same way contact of imidacloprid and clothianidin cause a lessening of foraging activity.

### 7.5 Foraging

Foraging behavior important to maintain the food supply in honeybee colony and so determine honeybee colony's survival and proliferation fortune. Repellent pesticides or anti-feedant usually modify bee foraging behavior, However, prolong exposure to it causes colony starvation and nutritional deficiency. Carbamate and Organophosphate pesticides disturb cholinesterase enzyme i.e. acetylcholinesterase, a vital enzyme control nerve impulse transmission. Organophophte fenitrothion causes an intense reduction in the foraging bees on flowering. When bees make acute contact with organophosphate, the toxicity effects are more intense [86]. Infield, a reduction in the food at visiting site could simply be elucidated in term of repellent effect. Neonicotinoid insecticides such as imidacloprid show a moderate toxic action [87]. Toxicity of imidacloprid comprised a reduction in the number foraging trips, less number of active bees at foraging sites, an enhanced in time intervals between visits, irregularities in communication or waggle dance, and disrupt in visual learning and navigatio. This demonstrates that the excessive stimulation of the acetylcholinesterase, inhibition of organophosphate or an agonist action on nAChR, with neonicotinoids [88, 89].

## 7.6 Muscle activity

Muscle contraction is important in almost all physiological functions and plays an important role in the accomplishment of tasks such as communication dance, flight, digestion of food, and heart and wing thrashing. The pesticides effects on the muscles of honeybees have been studied. In the honeybee research has been carried out on deltamethrin and prochloraz fungicide. In a laboratory study a semi-isolated heart exposed to prochloraz and deltamethrin diluted solutions showed a rapidly reduction in the frequency and the cardiac contractions force [90]. Results indicated that prochloraz seems more cardiotoxic than deltamethrin. Deltamethrin is a neurotoxic substance whereas prochloraz is an inhibitor and an inducer of cytochrome P450 enzymes, which are involved in the detoxication and metabolism of xenobiotics. An association of prochloraz and deltamethrin provokes synergistic effects that entirely and quickly stops heart muscle contractions, thus endorsing the neural basis of the deltamethrinprochloraz lethal and thermogenic synergy. Phenoxyacetic herbicide 2, 4-D also prompts parallel effects. Conversely, triazine herbicides subsidize tranquil neurotransmitter release that results in a boosted frequency and cardiac contractions force [91, 92]. Pyrethroids toxicity correlated with action on voltage-dependent sodium channels. In fact, pyrethroid pesticides and azole fungicides combine together and produce synergistic effects, and disturb normal functioning of target cells such as ATPase and potassium and calcium channels, that play a vital role in muscle and nerve activity [93, 94].

## 7.7 Thermoregulation

Thermoregulation is a phenomenon in honeybees to regulate their body temperature. Thermoregulation comprises thermogenesis by unremitting flight muscles contraction, fanning and beating wings, and evaporation of water at the individual as well as colony level. The thermoregulation allows the bees to fly at temperatures ranging from 11–46°C and heat their swarms and their broods. It is also plays a vital role in the exchange of information during foraging, waggle dnce, social communication and speedy nectar processing [95, 96]. Insecticides disturbed the normal thermoregulation process. Pyrethroid pesticides and azole fungicides (imidazoles and triazoles) usually act synergistically to prompt lethal toxicity. Azole fungicides act as both inhibitors and inducers of cytochromes P-450 enzymes involved in the metabolism of xenobiotics [97]. This initially suggested the action of azole fungicides on pyrethroid metabolism. Pyrethroids display a negative temperature coefficient below 28–30°C and induce higher toxicity at low temperatures. Honeybees were exposed to sub-lethal doses of deltamethrin to evaluate the toxicity. It was concluded that sub-lethal concentrations of deltamethrin triggers extreme hypothermia conditions for about 4 h [98]. Similar organophosphate also created thermogenesis process impairment by inhibiting acetylcholinesterase. These finding suggested association of a cholinergic pathway in the negative control of thermogenesis. Deltamethrin exposure induces hypothermia and leads to disruption of homing flight during foraging. The results of these study showed that condition of hypothermia created after impairment of muscle contraction rhythms and disturb nerve impulse process. Hence, honeybees would be capable of coordinated flight muscle contraction but would not be able to perform the neural program to attain shivering. Hypothermia become more sever in winter and spring when temperature already low and necessitating efficient thermogenesis [99].

### 7.8 Effects on reproduction

Reproduction is an important natural phenomenon that guarantee the existence and proliferation of the colony population. Indeed, a loss of reproductive brood consider as more detrimental for the bee colony than the loss of older bees (foragers). In hive, bees showed eusocial perform division of tasks to maintain the colony. Worker bees accomplish numerous chores throughout their lifespan. A few studies have revealed the adverse effects on larval development after exposure to pesticides [100]. It is evaluated a delay in hatching and development of honeybee when fed with imidacloprid contaminated food.

## 8. Minimizing honeybee exposure to pesticides

The threats to pollinators are significant and complex. Managing all threats in an integrated way will be an immense and fundamentally necessary task. Transforming the current chemical-intensive agriculture system into an ecological farming system will have much positive impact on the environment, pollinators, and human food security. However, these methods are often neglected as potentially effective tools for protecting the bee population. A recent study in Sweden clearly showed how strawberry crops benefited from organic farming. Organic strawberries received more pollinators and achieved higher pollination success than conventionally grown strawberries, and this difference was evident quickly after the conversion from conventional to organic farming. The authors concluded that organic agriculture benefited crop pollination in terms of both the quantity and quality of the yield [99]. Diversified farming systems, like those under organic or ecological production methods, bring out many benefits in addition to increased pollination services; they enhance the control of weeds, diseases, and insect pests. However, these systems have received significantly less public funding for research as a means of improved management, compared to conventional farming systems. This lack of support is remarkable, given that ecological and organic farming systems can produce approximately the same amount of food and profits as conventional farming while generating far less environmental and social harm [101]. More public and private funding are needed for research and development on ecological farming practices that enhanced ecological services, alongside food production and environmental protection, while at the same time helping social and economic development.

## 9. Food security crises due to honeybee decline

Scientists accepted that a world without honeybees would have a critical situation of food shortages and possibly leads to famine. There is much debate over whether Albert Einstein once said, that "if bees disappeared off the face of the Earth, humans would have only four years left to live. If all bees completely vanished, it would not cause humans to go extinct". Among all the edible crops that fulfill about 90% of food requirements throughout the world, among them 70% are pollinated by honeybees and other pollinators. Pollinators also improve the quality and shelf-life of crops and, also increase the genetic variability. It is estimated that the economic value of insect pollinating crops is about 761 € per ton [102]. Further, the majority of calories of human bodies also fulfill from insect-pollinated plants. The decline of the pollinator population cause decrease of a substantial amount of high economic values crops and fruits. These crops are a key element of the majority of calories and vitamins. In terms of nutrients in the human diet, they account for more than 90% of vitamin C, 100% of Lycopene and almost 100% of the antioxidants β-cryptoxanthin and β- tocopherol, the majority of the lipids (74%), vitamin A (>70%) and related carotenoids (98%), calcium (58%) and fluoride (62%), and a large portion of folic acid (55%). In total, pollinator-mediated crops account for about 40% of the global nutrient supply for humans [103]. Pollinators are essential to fulfill all nutrient requirements as they are responsible for essential crop pollination. If all the insect pollinators kill it would lead to a drastic decrease in fruit setting and growth of edible crops, which would affect all the population that is depending on it. It would also disturb the herbivores population that depends upon plants, which in turn disturb carnivores. In simple whole food chain will be disturb. Honeybees are under great threat due to the combined effects of global warming, intensive agriculture practices, habitat loss and, insecticide uses. FAO's Director-General said that the absence of honeybees and other pollinators will lead to wiping out coffee, orange, apple, peaches, tomatoes, and other cereal crops that rely upon pollination. Countries need to move toward more pollinator-friendly practices. It is estimated that only in North America 30% of consuming food is produced from bee pollinating vegetation. The value of bee pollination is estimated at $16 billion only in the USA. In the last few years particularly between 2007 to 2016 honeybee population size has been dramatically reduced, dropping by 89 percent. In the United States, 40 percent of honeybee colonies were lost in 2018 alone. In the USA annual income of fruits, seed, and nuts plants, corn crops decrease due to the decline or disappearance of 90% pollinator decline. In the USA extensive use of insecticides has been considered as a prominent factor for colony disappearance, bee population decline, honey production, and wax yield and cause losses of about $283 million per year. In Africa, the economic value of insect-pollinating crops is $11.6 million per year. In Pakistan yield of some fruit reduces up to 33.4% due to pollinator decline [104]. In the Himalayan region of Pakistan, the population of pollinators declines due to farmers' and institutions' unawareness about pollination benefits.

Globally, pollination has an estimated market value of up to $577 billion annually which represents about 10 percent of the global crop market [105]. Without these biotic pollinating services, decreases in crop production could both surge prices for consumers and lead to producers a loss of nearly $2 billion annually. It argues a future with compromised pollination due to the absence of pollinating insects points to a dominant urge for hand pollination or any other innovative technology. But this involved labor cost in terms of hand pollination or investment by innovative technology. The labor costs involved in hand pollination are potentially significant, estimated at $90 billion per year in the United States alone.

## 10. Conclusion and future recommendations

Pesticide exposure can induce more or less harmful effects on neural functions and cause disturbance of maintenance behavior and physiological functions. The mechanisms by which pesticides produce their lethal and sub-lethal effects are not limited to the exclusive interaction between the pesticide active substance and the target molecular. The effects of pesticides encompass several molecular targets sites of diverse affinities at various exposure and contact levels. Exposure time was also studied as a significant factor in insecticide and pesticide toxicity. The toxicity of pesticides is correlated to circadian rhythms, the exposure duration, the developmental stage of the bees exposed to pesticides, and the seasonal stress. The route and the mode of exposure (acute, sub-chronic, or chronic) play a predominantly decisive role in pesticide toxicity. Metabolic processes modulate the intrinsic toxicity and resultant metabolites may cause high or low toxicity than that of the parent pesticide. Researches on synergistic effects of pesticides should take a more dominant place in honeybee toxicology in the future. The practices of a large number of pesticides at a single site, synergistic effects by pesticide combinations will need to be prioritized. This could be based on the spatio-temporal presence of pesticide active substances. A promising approach to avoiding side effects of pesticides on pollinators has been derived by improving the mode of action of pesticides in targeted pests and honeybees. Mode of action can be improved by selecting the active substances that act selectively on pests [106].

With the extensive use of pesticides, the conservation of foraging honeybees is challenging. Even if a single worker of bee brings pesticide-contaminated pollen it can harm large numbers of honeybees. Systematic insecticides are a great threat to bees. Systemic insecticide disrupts the foraging behavior by impairing cholinergic signaling. Pesticide exposure also disturbs long and short-term memory. Insecticide treatments are not recommended in plantations and crops during flowering. Based on these findings it is recommended to apply methods of integrated pest management (IPM) of crops should be used, avoiding the use of chemical products. This report shows scientific evidence about the harmful effects of pesticides on honeybees and proving that pesticides play important role in bee decline. As consequence policymakers should ban the bee-harming pesticides. Through national action plans to support and promote agricultural practices such as crop rotation and organic farming that benefits the pollinator population. For this purpose promote research on ecological and organic farming practices that move away from reliance on chemical pest control.

## Author details

Farkhanda Manzoor and Mahnoor Pervez*
Department of Zoology, Lahore College for Women University, Lahore, Pakistan

*Address all correspondence to: mahnoorentomology@gmail.com

## References

[1] Velthuis HH, A century of advances in bumblebee domestication and the economic and environmental aspects of its commercialization for pollination. Apidologie, 2006; 37 (4): p. 421 - 451.

[2] Rockstrom J, Steffen W. A safe operating space for humanity. Nature, 2009; 461 (7263): 472 - 475.

[3] Gallai N, Salles JM, Settele J, Vaissiae BE. Economic valuation of the vulnerability of world agriculture confronted with pollinator decline. Ecological economics. 2009; **68** (3): 810 - 821.

[4] Biesmeijer JC, Roberts SPM, Reemer M, Ohlemuller R, Edwards M, Peeters T, Schaffers AP, Potts SG, Kleukers R, Thomas, CD, Settele J, Kunin W E. Parallel declines in pollinators and insect-pollinated plants in Britain and the Netherlands. Science, 2006; **313** (5785): 351 - 354.

[5] Stokstad E. The case of the empty hives. Science, 2007; **316** (5827):970 - 972.

[6] Aizen M A, Harder, LD. The global stock of domesticated honey bees is growing slower than agricultural demand for pollination. Current biology, 2009; **19**(11): p. 915-918.

[7] Lebuhn G, Droege S, Connor EF, Gemmill HB, Potts SG, Minckley RL, Griswold T, Jean R, Kula E, Roubik DW, Cane J, Wright K W, Frankie G Parker F. Detecting insect pollinator declines on regional and global scales. Conservation biology, 2013; **27**(1): p. 113-120.

[8] Potts SG, Biesmeijer JC, Kremen C, Neumann P, Schweiger O, Kunin WE. Global pollinator declines: trends, impacts and drivers. Trends in ecology and evolution, 2010; **25**(6): 345 - 353.

[9] Kluser S, Peter N, Chauzat MP, Pascal P, Ron W, Mwangi T. Global honey bee colony disorders and other threats to insect pollinators. 2010; 16: 1 - 16.

[10] Decourtye A, Devillers J, Aupinel P, Brun F, Bagnis C, Fourrier J, Gauthier M. Honeybee tracking with microchips: a new methodology to measure the effects of pesticides. Ecotoxicology, 2011; **20**(2): 429 - 437.

[11] Genersch E, Von der O W, Kaatz H, Schroeder A, Otten C, Bachler R, Berg S, Ritter W, Mohlen W, Gisder S, Meixner M, Liebig G and Rosenkranz P. The German bee monitoring project: a long term study to understand periodically high winter losses of honey bee colonies. Apidologie, 2010; **41**(3): 332 - 352.

[12] Higes M, Martin R, Meana A. *Nosema ceranae*, a new microsporidian parasite in honeybees in Europe. Journal of invertebrate pathology. 2006; **92**(2): 93 - 95.

[13] Memmott J, Craze P G, Waser N M, Price M V. Global warming and the disruption of plant–pollinator interactions. Ecology letters, 2007; **10**(8): 710 - 717.

[14] Rabea E I, Nasr, HM Badawy ME. Toxic effect and biochemical study of chlorfluazuron, oxymatrine, and spinosad on honey bees (*Apis mellifera*). Archives of environmental contamination and toxicology, 2010; **58**(3): 722 - 732.

[15] Nauen RU, Ebbinghaus-Kintscher, Schmuck, R. Toxicity and nicotinic acetylcholine receptor interaction of imidacloprid and its metabolites in *Apis mellifera* (Hymenoptera: Apidae). Pest Management Science: formerly Pesticide Science, 2001; **57**(7): 577 - 586.

[16] Bonmatin J M, Moineau I, Charvet R, Fleche C, Colin M E, Bengsch E R. A LC/APCI-MS/MS method for analysis of imidacloprid in soils, in plants, and in

pollens. Analytical chemistry, 2003; **75**(9): 2027 - 2033.

[17] Bonmatin J M, Giorio C, Girolami V, Goulson D, Kreutzweiser D P, Krupke C. Environmental fate and exposure; neonicotinoids and fipronil. Environmental science and pollution research, 2015; **22**(1): 35 - 67.

[18] Decourtye A, Lacassie E, Pham-Delegue M H. Learning performances of honeybees (*Apis mellifera* L) are differentially affected by imidacloprid according to the season. Pest Management Science, 2003; **59** (3): 269 - 278.

[19] Wallner K. Sprayed and seed dressed pesticides in pollen, nectar and honey of oilseed rape. Julius-Kühn-Archiv, 2009; (423): 152 - 153.

[20] Rortais A, Arnold G, Halm M P, Touffet B F. Modes of honeybees exposure to systemic insecticides: estimated amounts of contaminated pollen and nectar consumed by different categories of bees. Apidologie, 2005; **36** (1): 71 - 83.

[21] Greatti M, Barbattini R, Stravisi A, Sabatini A G, Rossi S. Presence of the ai imidacloprid on vegetation near corn fields sown with Gaucho® dressed seeds. Bulletin of insectology, 2006. **59**(2): p. 99.

[22] Tapparo A, Marton D, Giorio C. Assessment of the environmental exposure of honeybees to particulate matter containing neonicotinoid insecticides coming from corn coated seeds. Environmental science and technology, 2012; **46**(5): 2592 -2599.

[23] Skerl M I S, Bolta S V, Cesnik H B, Gregorc A. Residues of pesticides in honeybee (*Apis mellifera* carnica) bee bread and in pollen loads from treated apple orchards. Bulletin of environmental contamination and toxicology, 2009; **83**(3): 374 - 377.

[24] Chauzat M P, Faucon J P, Martel A C, Lachaize J, Cougoule N, Aubert M A A survey of pesticide residues in pollen loads collected by honey bees in France. Journal of economic entomology, 2006; **99**(2): 253 - 262.

[25] Krupke C H, Hunt G J, Eitzer B D, Andino G, Given K. Multiple routes of pesticide exposure for honeybees living near agricultural fields. PLoS One, 2012; **7**(1): e29268.

[26] Bernal J, Garrido BE, Nozal MJD, Gonzalez PAG, Hernandez MR, Diego JC, Jimenez JJ, Bernal JL, Higes M. Overview of pesticide residues in stored pollen and their potential effect on bee colony (*Apis mellifera*) losses in Spain. Journal of economic entomology, 2010; **103**(6): 1964 - 1971.

[27] Girolami V, Mazzon L, Squartini A. Translocation of neonicotinoid insecticides from coated seeds to seedling guttation drops: a novel way of intoxication for bees. Journal of economic entomology, 2009; **102** (5): 1808 - 1815.

[28] Orantes B F J, Pajuelo A, Megias M, Torres F P C. Pesticide residues in beeswax and beebread samples collected from honey bee colonies (*Apis mellifera* L.) in Spain. Possible implications for bee losses. Journal of apicultural research, 2010; **49**(3): 243 - 250.

[29] Mullin C A, Frazier M, Frazier J L, Ashcraft S, Simonds R and Pettis JS. High levels of miticides and agrochemicals in North American apiaries: implications for honey bee health. Plos One, 2010; **5**(3): e9754.

[30] Henry M, Beguin M, Fabrice R, Orianne R, Jean O, Pierrick A, Jean A, Sylvie T, Axel D A. Response to comment on "A common pesticide decreases foraging success and survival in honeybees". science, 2012; **337**(6101): 1453 - 1453.

[31] Chakrabarti P, Rana S, Sarkar S, Smith B, Basu P. Pesticide-induced oxidative stress in laboratory and field

populations of native honeybees along intensive agricultural landscapes in two Eastern Indian states. Apidologie, 2015; **46**(1): 107 - 129.

[32] Zaluski R, Kadri S M, Alonso D P, Ribolla P E, Oliveira O R. Fipronil promotes motor and behavioral changes in honey bees (*Apis mellifera*) and affects the development of colonies exposed to sublethal doses. Environmental toxicology and chemistry, 2015; **34**(5): 1062 - 1069.

[33] Iwasa T, Motoyama N, Ambrose JT, Roe RM. Mechanism for the differential toxicity of neonicotinoid insecticides in the honey bee. Crop protection, 2004; **23**(5): 371 - 378.

[34] Decourtye A, Devillers J, Genecque E, Le Menach K, Budzinski H, Cluzeau S, Pham-Delegue MH. Comparative sublethal toxicity of nine pesticides on olfactory learning performances of the honeybee *Apis mellifera*. Archives of environmental contamination and toxicology, 2005; **48**(2): 242 - 250.

[35] Desneux N, Decourtye A, Delpuech, JM. The sublethal effects of pesticides on beneficial arthropods. Annual Review of Entomology 2007; **52**: 81 - 106.

[36] Nauen R, Denholm I. Resistance of insect pests to neonicotinoid insecticides: current status and future prospects. Archives of insect biochemistry and physiology: Published in Collaboration with the Entomological Society of America, 2005; **58**(4): 200 - 215.

[37] El Hassani A K, Dupuis JP, Gauthier M, Armengaud C. Glutamatergic and GABAergic effects of fipronil on olfactory learning and memory in the honeybee. Invertebrate neuroscience, 2009; **9**(2): 91.

[38] Schmidt H W, Brasse D, Kunast C, Muhlen W, Vonder OW, Tornier I, Wallner K. Introduction of indices for the evaluation of tent tests and field tests with honeybees. Bulletin of insectology, 2003; **56**: 111 - 118.

[39] Laurino D, Porporato M, Patetta A, Manino A. *Toxicity of neonicotinoid insecticides to honeybees: laboratory tests.* Bulletin insectology, 2011; **64**(1): 107 - 13.

[40] Elbert C, Erdelen C, Kuehnhold J, Nauen R, Schmidt H W, Hattori Y. Thiacloprid, a novel neonicotinoid insecticide for foliar application. in The BCPC Conference: Pests and diseases, Volume 1. Proceedings of an international conference held at the Brighton Hilton Metropole Hotel, Brighton, UK, 13 - 16 November 2000. British Crop Protection Council.

[41] Rondeau G, Bayo S F, Tennekes H A, Decourtye A, Romero R, Desneux, N. Delayed and time-cumulative toxicity of imidacloprid in bees, ants and termites. Scientific reports, 2014; **4**(1): 1 - 8.

[42] Pettis J S, Johnson J, Dively G. Pesticide exposure in honey bees results in increased levels of the gut pathogen Nosema. Naturwissenschaften, 2012; **99**(2): 153 - 158.

[43] Chauzat M P, Martel A C, Cougoule N, Porta P, Lachaize J, Zeggane S, Aubert M, Caroentier P, Faucon J P. An assessment of honeybee colony matrices, *Apis mellifera* (Hymenoptera: Apidae) to monitor pesticide presence in continental France. Environmental toxicology and chemistry, 2011; **30**(1): 103 - 111.

[44] Kiljanek T, Alicja N, Stanisław S, Marta G, Milena B, Andrzej P. Multi-residue method for the determination of pesticides and pesticide metabolites in honeybees by liquid and gas chromatography coupled with tandem mass spectrometry—Honeybee poisoning incidents. Journal of chromatography, 2016; **1435**: 100 - 114.

[45] Bortolotti L, Montanari R, Marcelino J, Medrzycki P, Maini S,

Porrini C. Effects of sub-lethal imidacloprid doses on the homing rate and foraging activity of honey bees. Bulletin of insectology, 2003; **56**: 63 - 68.

[46] Porrini C, Sabatini A G, Girotti S, Fini F, Monaco L, Celli G, Bortolotti L, Ghini, S. The death of honey bees and environmental pollution by pesticides: the honey bees as biological indicators. Bulletin of insectology, 2003; **56**(1): 147 - 152.

[47] Oliveira RA, Roat TC, Carvalho SM Malaspina O. Side-effects of thia-methoxam on the brain andmidgut of the africanized honeybee *Apis mellifera* (Hymenopptera: Apidae). Environmental toxicology, 2014; **29**(10): 1122 - 1133.

[48] Suchail S, Guez D, Belzunces LP. Discrepancy between acute and chronic toxicity induced by imidacloprid and its metabolites in *Apis mellifera*. Environmental Toxicology and Chemistry, 2001; **20**(11): 2482 - 2486.

[49] Lambin M. Imidacloprid-induced facilitation of the proboscis extension reflex habituation in the honeybee. Archives of insect biochemistry and physiology: Published in Collaboration with the Entomological Society of America, 2001; **48**(3): 129 - 134.

[50] Van dame R, Meled M, Colin MA. Alteration of the homing-flight in the honey bee *Apis mellifera* L. Exposed to sublethal dose of deltamethrin. Environmental toxicology and chemistry, 1995; **14**(5): 855 - 860.

[51] Easton AH, Goulson D. The neonicotinoid insecticide imidacloprid repels pollinating flies and beetles at field-realistic concentrations. PLoS One, 2013; **8**(1): e54819.

[52] Whitehorn PR, Connor S, Wackers FL, Goulson D. Neonicotinoid pesticide reduces bumble bee colony growth and queen production. Science, 2012; **336**(6079): 351 - 352.

[53] Levot GW, An insecticidal refuge trap to control adult small hive beetle, *Aethina tumida* Murray (Coleoptera: Nitidulidae) in honeybee colonies. Journal of apicultural research, 2008; **47**(3): 222 - 228.

[54] Esch, HE, Zhang SW, Srinivasan MV, Tautz J. Honeybee dances communicate distances measured by optic flow. Nature, 2001; **411**(6837): 581 - 583.

[55] Cresswell JE. A meta-analysis of experiments testing the effects of a neonicotinoid insecticide (imidacloprid) on honeybees. Ecotoxicology, 2011; **20**(1): 149 - 157.

[56] Tomizawa M, Casida JE, Neon-icotinoid insecticide toxicology: mechanisms of selective action. Annual review of pharmacology and toxicology. 2005; **45**: 247 -268.

[57] Casida J E, Durkin K A. Neuroactive insecticides: targets, selectivity, resistance, and secondary effects. Annual review of entomology, 2013; **58**: 99 - 117.

[58] Hardstone MC, Scott JG. Is *Apis mellifera* more sensitive to insecticides than other insects? Pest management science, 2010; **66**(11): 1171 - 1180.

[59] Barnett E A, Charlton AJ, Fletcher MR. Incidents of bee poisoning with pesticides in the United Kingdom, 1994 - 2003. Pest management science, 2007; **63**(11): 1051 - 1057.

[60] Johnson R M, Ellis M D, Mullin C A, Frazier M. Pesticides and honeybee toxicity–USA. Apidologie, 2010; **41**(3): 312 - 331.

[61] Chauzat M P. Faucon JP. Pesticide residues in beeswax samples collected from honey bee colonies (*Apis mellifera* L.) in France. Pest Management Science: formerly Pesticide science, 2007; **63**(11): 1100 - 1106.

[62] Haarmann T, Spivak M, Weaver D, Weaver B, Glenn T. 2002. Effects of fluvalinate and coumaphos on queen honey bees (Hymenoptera: Apidae) in two commercial queen rearing operations. Journal of economic entomology, 2002; **95**(1): 28 - 35.

[63] Guez D, Belzunces L, Maleszka R. Effects of imidacloprid metabolites on habituation in honeybees suggest the existence of two subtypes of nicotinic receptors differentially expressed during adult development. Pharmacology biochemistry and behavior, 2003; **75**(1): 217 - 222.

[64] Guez D, Suchail S, Gauthier M, Maleszka R, Belzunces L P. Contrasting effects of imidacloprid on habituation in 7-and 8-day-old honeybees (*Apis mellifera*). Neurobiology of learning and memory, 2001; **76**(2): 183 - 191.

[65] Moore D. Honey bee circadian clocks: behavioral control from individual workers to whole-colony rhythms. Journal of insect physiology, 2001. **47** (8): 843 - 857.

[66] Guez D, Zhu H, Zhang S W, Srinivasan M V. Enhanced cholinergic transmission promotes recall in honeybees. Journal of insect physiology, 2010; **56**(9): 1341 - 1348.

[67] Decourtye A, Devillers J, Cluzeau S, Charreton M, Pham-Delegue M H. Imidacloprid impairs memory and brain metabolism in the honeybee (*Apis mellifera* L.). Pesticide biochemistry and physiology, 2004; **78**(2): 83 - 92.

[68] Han P, Niu C Y, Lei C L, Cui J J, Desneux N. Use of an innovative T-tube maze assay and the proboscis extension response assay to assess sublethal effects of GM products and pesticides on learning capacity of the honeybee *Apis mellifera* L. Ecotoxicology, 2010; **19**(8):1612 - 1619.

[69] Goulson D, Chapman J W, HughesW O. Discrimination of unrewarding flowers by bees; direct detection of rewards and use of repellent scent marks. Journal of insect behavior, 2001; **14**(5): 669 - 678.

[70] Aliouane Y, Adessalam K, El Hassani A K, Gray V, Armengaud C, Lambin M, Gauthier M. Subchronic exposure of honeybees to sublethal doses of pesticides: effects on behavior. Environmental Toxicology and Chemistry: An international Journal, 2009; **28**(1):113 - 122.

[71] Bernadou A, Demares F, Couret-Fauvel T, Sandoz J C, Gauthier M. Effect of fipronil on side-specific antennal tactile learning in the honeybee. Journal of insect physiology, 2009; **55**(12): 1099 – 1106.

[72] Reinhard J, Srinivasan M V, Zhang S. Scent-triggered navigation in honeybees. Nature, 2004; **427** (6973) 411 - 411.

[73] Gawleta N, Zimmermann Y, Eltz T. Repellent foraging scent recognition across bee families. Apidologie, 2005; **36**(3): 325 - 330.

[74] Couvillon M J. Alarm pheromones do not mediate rapid shifts in honeybee guard acceptance threshold. Journal of chemical ecology, 2010; **36**(12): 1306-1308.

[75] Kather R, Drijfhout F P, Martin S J. Task group differences in cuticular lipids in the honeybee *Apis mellifera*. Journal of chemical ecology, 2011; **37**(2): 205 - 212.

[76] El Hassani A K, Dacher M, Gary V, Lambin M, Gauthier M, Armengaud C. Effects of sublethal doses of acetamiprid and thiamethoxam on the behavior of the honeybee (*Apis mellifera*). Archives of environmental contamination and toxicology, 2008; **54**(4): 653 - 661.

[77] Barron A B, Zhu H, Robinson G E, Srinivasan M V. Influence of flight time and flight environment on distance

communication by dancing honey bees. Insectes sociaux, 2005; **52**(4): 402 - 407.

[78] Menzel R, De Marco R J, Greggers U. Spatial memory, navigation and dance behaviour in *Apis mellifera*. Journal of comparative physiology, 2006; **192**(9): 889 - 903.

[79] Dacke M, Srinivasan M V. Honeybee navigation: distance estimation in the third dimension. Journal of Experimental Biology, 2007; **210** (5): 845 - 853.

[80] Reynolds D, Riley J. Remote-sensing, telemetric and computer-based technologies for investigating insect movement: a survey of existing and potential techniques. Computers and electronics in agriculture, 2002; **35**(2): 271 - 307.

[81] Riley J R, Greggers U, Smith A D, Reynolds D R, Menzel R. The flight paths of honeybees recruited by the waggle dance. Nature, 2005; **435** (7039): 205 - 207.

[82] Capaldi E A, Smith A D, Osborne J L, Fahrbach S E, Farris S M, Reynolds D R, Edwards A S, Martin A, Robinson G E, Poppy G M, Riley J R. Ontogeny of orientation flight in the honeybee revealed by harmonic radar. Nature, 2000; **403**(6769): 537 - 540.

[83] Menzel R, Greggers U, Smith A, Berger S, Brandt R, Brunke S, Bundrock G, Hulse S, Plumpe T, Schaupp F, Schuttler E, Stach S, Stindt J, Stollhoff N, Watzl S. *Honey bees navigate according to a map-like spatial memory.* Proceedings of the National Academy of Sciences, 2005; **102**(8): 3040 - 3045.

[84] Streit S, Bock F, Pirk CWW, Tautz J. *Automatic life-long monitoring of individual insect behaviour now possible.* Zoology, 2003; **106**(3): 169 - 171.

[85] Ohashi K, Souza D D, Thomson J. *An automated system for tracking and identifying individual nectar foragers at multiple feeders.* Behavioral Ecology and Sociobiology, 2010; **64**(5): 891 - 897.

[86] Guez D, Zhang S.W, Srinivasan M.V. *Methyl parathion modifies foraging behaviour in honeybees (Apis mellifera).* Ecotoxicology, 2005; **4**(4): 431 - 437.

[87] Colin M E, Bonmatin J M, Moineau I, Gaimon C, Brun S, Vermandere JP*A method to quantify and analyze the foraging activity of honey bees: relevance to the sublethal effects induced by systemic insecticides.* Archives of environmental contamination and toxicology, 2004; **47**(3): 387 - 395.

[88] Ramirez R, Chaufaux J, Pham-Delegue MH. *Effects of Cry1Ab protoxin, deltamethrin and imidacloprid on the foraging activity and the learning performances of the honeybee Apis mellifera, a comparative approach.* Apidologie, 2005; **36**(4): 601 - 611.

[89] Yang E C, Chuang Y C, Chen Y L, Chang L H. *Abnormal foraging behavior induced by sublethal dosage of imidacloprid in the honey bee (Hymenoptera: Apidae).* Journal of economic entomology, 2008; **101**(6): 1743 - 1748.

[90] Papaefthimiou C G. Theophilidis, *The cardiotoxic action of the pyrethroid insecticide deltamethrin, the azole fungicide prochloraz, and their synergy on the semi-isolated heart of the bee Apis mellifera macedonica.* Pesticide biochemistry and physiology, 2001; **69**(2): 77 - 91.

[91] Papaefthimiou C, Theophilidis G. *Triazines facilitate neurotransmitter release of synaptic terminals located in hearts of frog (Rana ridibunda) and honeybee (Apis mellifera) and in the ventral nerve cord of a beetle (Tenebrio molitor).* Comparative biochemistry and physiology Part C: Toxicology & Pharmacology, 2003; **135**(3): 315 - 330.

[92] Soderlund DM, Clark JM, Sheets LP, Mullin LS, Piccirillo VJ, Sargent D, Stevens JT, Weiner ML. Mechanisms of

pyrethroid neurotoxicity: implications for cumulative risk assessment. Toxicology, 2002; **171**(1): 3 - 59.

[93] Calore EE, Cavaliere MJ, Puga FR, Calore NMP, Rosa A R, Weg R, Dias S S, Santos R P. Histologic peripheral nerve changes in rats induced by deltamethrin. Ecotoxicology and environmental safety, 2000; **47**(1): 82 - 86.

[94] Neal AP, Yuan Y, Atchison WD. Allethrin differentially modulates voltage-gated calcium channel subtypes in rat PC12 cells. Toxicological sciences, 2010; **116**(2): 604 - 613.

[95] Stabentheiner A, Vollmann J, Kovac H, Crailsheim K. Oxygen consumption and body temperature of active and resting honeybees. Journal of insect physiology, 2003; **49**(9): 881 - 889.

[96] Kleinhenz, M., Bujok, B., Fuchs, S., Tautz, J. Hot bees in empty broodnest cells: heating from within. Journal of experimental biology, 2003; **206**(23): 4217 - 4231.

[97] Babin M, Casado S, Chana A, Herradon, B., Segner, H., Tarazona, J.V., Navas, J.M. Cytochrome P4501A induction caused by the imidazole derivative Prochloraz in a rainbow trout cell line. Toxicology in vitro, 2005; **19**(7): 899 - 902.

[98] Vandame R, Belzunces LP. Joint actions of deltamethrin and azole fungicides on honey bee thermo-regulation. Neuroscience letters, 1998; **251**(1): 57 - 60.

[99] Andersson GK, Rundlof M, Smith HG. Organic farming improves pollination success in strawberries. PloS one, 2012; 7(2):e31599.

[100] Tasei JN. Effects of insect growth regulators on honey bees and non-Apis bees. A review. Apidologie, 2001; **32**: 527 - 545.

[101] Davis AS, Hill JD, Chase CA, Johanns AM, Liebman M. Increasing cropping system diversity balances productivity, profitability and environmental health. PloS One, 2012. 7(10): p. e47149.

[102] Chagnon M. Risks of large-scale use of systemic insecticides to ecosystem functioning and services. Environmental science and pollution research, 2015; **22**(1): 119 - 134.

[103] Eilers EJ, Claire K, Sarah SG, Andrea KG Alexandra MK. Contribution of pollinator-mediated crops to nutrients in the human food supply. PLoS One, 2011; **6**(6): e21363.

[104] Aslam M, Razzaq M, Hussain S Akhter W. Effect of insect pollinators on fruit setting on Mango (*Mangifera indica* L.). Journal of Science Research, 2004; **15**: 53 - 58.

[105] Chaplin-Kramer R, Dombeck E, Gerber J, Knuth K A, Mueller ND, Mueller M, Ziv G, Klein AM. Global malnutrition overlaps with pollinator-dependent micronutrient production. Proceedings of the Royal Society B: Biological Sciences, 2014; **281**(1794): 20141799.

[106] Courjaret R, Grolleau F, Lapied B. Two distinct calcium-sensitive and-insensitive PKC up-and down-regulate an α-bungarotoxin-resistant nAChR1 in insect neurosecretory cells (DUM neurons). European journal of neuroscience, 2003; **17**(10): 2023 - 2034.

Chapter 5

# Diversity, Importance and Decline of Pollinating Insects in Present Era

*Navkiran Kaur and Amritpal Singh Kaleka*

## Abstract

Pollination is a multi-million-year-old co-evolutionary process involving flowering plants and pollinators. It is one of the most important mechanisms in preservation and promotion of biodiversity as well as life on Earth. Pollinator diversity is essential for maintaining overall biological diversity in many habitats including agro-ecosystems. Pollinators are responsible for assisting reproduction in over 80% of the world's flowering plants. In their absence, humans and wildlife would go hungry. Insects are the most efficient pollinators as they play a crucial part in pollination ecology. Pollinators and their habitats have ecological, economic, cultural and social benefits. Pollination efficiency is highly dependent on certain attributes and characteristics of pollinators such as vision, anatomy, food preferences, olfaction, behaviour and learning ability. With the rapid growth of human population, our demand for food has also risen. Our agricultural systems will need to produce more food in a sustainable manner in the future to cope with this. Pollinators play an important role in these ecosystems and will continue to do so in the future. Because pollinators are so important to agriculture, we need to learn more about which crops require specific pollinators and how to best maintain and promote both wild and controlled species. Their diversity needs protection because there are specific relationships between certain crops and pollinators. Pollinator communities are suffering as a result of man-made habitat disruptions, including severe biodiversity loss. This diversity must be protected by combining conservation measures with sustainable farming practices which could increase crop yields while protecting insect pollinator species.

**Keywords:** Insects, Pollinators, Species, Crops, Diversity

## 1. Introduction

Pollination is a multi-million-year-old ecosystem process from which both flowering plants and pollinators get benefitted. Pollinating animals come to flowers for a variety of reasons, including food and shelter. Pollen rubs or falls onto pollinator's bodies when they visit flowers. As the pollinator passes from one flower to the next, it transfers the pollen to another flower. This transfer is important in the life cycle of all flowering plants because it is required to begin seed and fruit production. Pollinators are important for healthy, productive agricultural ecosystems and nature.

Indeed, the interactions between plants and their pollinators are among the most beautiful examples of coevolution on the planet. While some pollinators are generalists, visiting a wide variety of flowers, many pollinators have acquired preferences for certain flower kinds, and vice versa. Most pollinators have their favourite colour of flower: Bees prefer blue flowers, butterflies prefer pink and red flowers, flies choose yellow and white flowers, beetles and bats prefer white flowers, while hummingbirds prefer red flowers. In addition, the phenology, form, and food reward offered by the flower can all impact which pollinators visit [1]. Bees, for example, can see ultraviolet light and have a better sense of bilateral symmetry. As a result, flowers that want to attract bees will likely use these visual signals to lure the bee to the flower's centre [2].

Though some plant species depend on wind or water currents to carry pollens from one flower to another, but majority of plant species (approx. 90%) prefer animal assistance in this task. Around 200,000 different species of animals do this task of pollen transfer. Out of these, 1,000 are of vertebrates (birds, bats and tiny mammals), with the remainder being invertebrates, such as moths, bees, flies, beetles and butterflies [3].

Plant-pollinator interactions may be one of the most ecologically significant types of animal–plant interactions: without pollinators, many plants would be unable to set seed or reproduce, and without plants to provide pollen, nectar, and other rewards, countless animal populations would decline, with knock-on effects for other species [4].

Plants and their pollinators have had a significant impact on each other's growth, frequently leading to diversification and even an exclusive partnership. The Madagascar Star Orchid (*Angraecum sesquipedale*), which possesses a foot-long nectar tube that can only be pollinated by a species of hawk moth (*Xanthopan morganii praedicta*) with its 8–14 inch long proboscis, is a good illustration of this. The exquisite Star Orchid-Hawk Moth relationship even helped Darwin supporting his theory of evolution [2].

## 2. Pollinator diversity

Mutualisms between plants and pollinators extend back to the Cretaceous period, when insects began to feed on flowers and flowers achieved higher reproductive success through the transfer of pollen by insects. At least 67 percent of blooming plants rely on insects for pollination today [5], with the rest relying on birds and mammals. Pollinators are just as important as light and water for these plants to survive [6].

Pollinators comprise a diverse group of animals that include species of butterflies, flies, moths, wasps, beetles, ants, birds, weevils, thrips, midges, bats, monkeys, marsupials, rodents, and reptiles, but are dominated by insects, particularly bees. Bees and flies visit more than 90% of the world's major plant types, while the other species visit fewer than 6% of the crop varieties (**Table 1**). The western and eastern species of honey bees i.e., *Apis mellifera* and *Apis cerana*, as well as some bumble bees, stingless bees and a few solitary bees are managed and the vast majority of world's known species of bees (20,077 species) are wild in nature, i.e., free-living and unmanaged [7].

Many species of flower visitors have been reported to visit flowering crops in the literature. For instance, a mega-study that included 90 percent of all agricultural pollination studies from throughout the world discovered that 785 different bee species visit crop blooms [8]. Bees are the most prolific and diverse pollinators in most parts of the world, with over 20,000 species recorded [9, 10]. With over 1,20,000

| Sr. No. | Pollinator group | Species name |
|---|---|---|
| 1. | Bumble bees | *Bombus affinis* Cresson, *Bombus californicus* Smith, *Bombus hortorum* Linnaeus, *Bombus hypnorum* Linnaeus, *Bombus impatiens* Cresson, *Bombus lapidarius* Linnaeus, *Bombus (Thoracobombus) pascuorum* Scopoli, *Bombus sonorus* Linnaeus, *Bombus terrestris* Linnaeus and *Bombus vosnesenskii* Radoszkowski |
| 2. | Beetles | *Carpophilus hemipterus* Linnaeus and *Carpophilus mutilatus* Erichson |
| 3. | Honey bees | *Apis cerana* Fabricius, *Apis dorsata* Fabricius, *Apis florea* Fabricius and *Apis mellifera* Linnaeus |
| 4. | Hover flies | *Trichometallea pollinosa* Townsend, *Eristalis cerealis* Fabricius and *Eristalis tenax* Linnaeus |
| 5. | Stingless bees | *Nannotrigona testaceicornis* Lepeletier, *Melipona favosa* Fabricius, *Melipona subnitida* Ducke, *Nanotrigona perilampoides* Cresson, *Trigona cupira* Smith, *Tetragonula iridipennis* Smith, *Tetragonula (Lepidotrigona) terminata* Smith, *Tetragonula (Tetragonoula) minangkabau* Sakagami, and *Scaptotrigona depilis* Moure |
| 6. | Thrips | *Thrips hawaiiensis* Morgan and *Haplothrips (Haplothrips) tenuipennis* Bagnall |
| 7. | Wasps | *Blastophaga psenes* Linnaeus |

**Table 1.**
*Species list of known pollinators for global crop.*

species, flies are an important group in agriculture, although only a few families are effective pollinators [11]. In colder climates, such as high altitude/latitude environments, flies outweigh bees in both diversity and quantity as pollinators [12]. In addition to bees and flies, butterflies, beetles, moths, wasps, ants, thrips and vertebrates also pollinate plants, including some crops. Pollinating butterflies and moths are found all around the planet, but in the tropics they are more numerous and diversified [13]. The enormous variety of insect pollinators was discussed by Kevan and Baker [14]. Some birds and bats, in addition to insects, are essential pollinators [15, 16]. Bird pollinators are mostly found in warm (tropical/subtropical) climates, whereas bats pollinate tropical forests and some desert cactus. Pollinators that are less well-known have also been reported for a variety of plant species. These include, among others, cockroaches [17], mice [18], squirrels [19], lizards [20–22] and snails [23]. The less well known pollinators are not known to have major roles in supporting agricultural production.

## 2.1 Bees

Bees play a significant role in pollination in most terrestrial environments around the world. Honeybees and thousands of species of native bees pollinate garden crops, meadows and woodland plants in the United States. The majority of bees visit flowers in search of pollen or nectar to nourish themselves and their young ones. Crop pollination and honey production are significantly reliant on honeybees. Solitary bees are among the most common native pollinators and named because most of them live solitary lives and do not assemble to live in colonies. Blueberries, sunflowers, apples, watermelon, alfalfa and strawberries are among the commercial crops pollinated by solitary bees. Solitary bees build their nests in a variety of unusual locations, such as sticks, mud mounds, and termite holes. A few species build mud nests and saps, plant resins on the edge of rocks and trees to make domed nests. Many bees excavate their nests into the soft inner pith of stems and twigs, or exploit abandoned beetle burrows. Some solitary bees, on the other hand, create tunnels in bare or partially vegetated, well-drained soil to make their nests. These bees can be generalist or specialist feeders, depending on the species. Generalist bees visit a wide variety of floral types collect nectar and pollens. Being more hardy species, these are able to thrive in degraded settings dominated by weedy or invasive plants. While specialists are more vulnerable to the detrimental effects of landscape or habitat changes since they depend on a single plant species for nectar and pollen.

Bumblebees are social bees, which means these bees reside in colonies, share tasks, and have many generations that overlap in the spring, summer, and fall. The bumblebees require a suitable sized cavity in to build their nest. These bees usually build their nest underground in abandoned rat burrows and sometimes in hollow trees or walls or under a clump of grass above ground. Bumblebees usually feed on a wide variety of plants.

### 2.2 Ants

Ants are gregarious insects that enjoy nectar in large quantities. These active insects are frequently seen visiting flowers in search of energy-dense nectar. Ants do not have any wings, so they have to crawl into each bloom to get their meal. They are more likely to collect nectar from flowers that are not efficiently cross-pollinated. Ants are drawn to low-growing, inconspicuous blooms close to the stem. Small's stonecrop (*Diamorpha smallii* Britton), alpine nailwort (*Paronychia pulvinata* Grey) and Cascade knotweed are examples of ant-pollinated plants in North America (*Polygonum cascadense* Baker).

### 2.3 Butterflies

Butterflies, like all pollinators, are inextricably related to their surroundings, and abrupt changes in the ecosystem can have fatal consequences for localised populations or species. The butterfly's habitat requirements differ from stage to stage, and each has its own set of requirements that must be taken into account in order to create acceptable habitat. The life cycle of a butterfly is divided into four stages: egg, caterpillar, pupa, and adult. Butterfly deposit its eggs on leaves of trees and shrubs, flowers and grasses.

Being oligolectic, most butterfly species remain confined to one or a few closely related species of plants as these plant species effectively act as host plants for their caterpillars. The females usually lay their eggs on or near the host plant for the survival of their caterpillars. The caterpillars of monarch butterflies, for example, only consume milkweed, and adult females of monarch butterflies lay eggs on or near milkweed plants. Newly hatched caterpillars feed on the leaves, stalks, flowers and fruits of their host plants, which also act as a protective barrier against predators. Caterpillars begin to transform into adult forms after several weeks of eating and growing. This is the pupal stage of a butterfly's life, which is a non-feeding, sedentary stage. Pupae do not require nourishment, but they do require a safe place to convert into their adult forms, such as sticks, tall grass or a pile of leaves.

Adult butterflies feed almost entirely on nectar. Butterflies prefer flowers that are brightly coloured, aromatic, and have flat, broad surfaces on which to land. Adult butterflies like the nectar of daisies such as zinnias, asters, marigolds, goldenrods, dahlias and asters, dogbane, butterfly weed, ironweed, phlox and milkweed. Rotting fruit, tree sap, mud puddles, animal excrement and urine are also sources of nutrients, minerals and salt for adult males of some species. Adult butterflies can feed, bask, and rest on the leaves and stems of the host plants, which provide perching locations. Wind, rain and predators can all be protected by vegetation and modest woodpiles.

### 2.4 Moths

The moths are nocturnal in nature and some species are pollinators of night-blooming flowering plants, especially in the southern United States and Mexico. The female yucca moth, for example, has mouthparts that allow her to capture pollen and lay her eggs in the stigma of the yucca flower. The life and propagation of yucca plants are entirely dependent on the yucca moth. Each flower's pistil (female component) terminates in a three-lobed stigma. Pollen masses must be driven down into this centre stigmatic opening in order for pollination to occur. Using her particularly modified mouthparts, the female yucca moth collects pollen from flower anthers. She gathers the sticky pollen and rolls it into a ball. She then "stuffs" or "combs" the pollen ball into the stigmas of the flowers she visits. The yucca flower will not develop into a fruit or pod with seeds unless this procedure occurs.

When a female moth visits a flower, she walks up to the base of the flower and inserts her ovipositor into one or more of the six chambers to lay an egg. The egg is protected in the chamber while it develops. The yucca will have begun to grow a pod with little seeds by the time the egg hatches into a tiny caterpillar. In this association, both the yucca plant and the yucca moth benefit.

### 2.5 Flies and beetles

Flies and beetles are two important pollinator groups. Certain species of flies show resemblance with bees by mimicking bee coloration and patterns. Both bees and flies possess transparent membranous wings and but flies can be distinguished on the basis of having only one pair of wings. Some pollinating beetles are small in size and difficult to spot as these beetles resemble with the black specks present on the petals of flowers, while others are large and more colourful. There are hundreds of thousands of species of pollinating flies and beetles, many of which have yet to be documented. The habitat requirements of different species vary. For each of their life phases, such as egg, larva, pupa, and adult, flies and beetles require food, water, and cover in adequate quantity and quality. Pollination is greatly aided by syrphid flies.

### 2.6 Wasps

Wasps, like bees, have extremely high energy requirements that must be satisfied in order for them to survive. Pollen and nectar from a variety of flowers are vital for wasps. True wasps have stingers, which they utilise to catch insects or spiders for their larvae to feed. Small fig wasps are common throughout the tropics. Many tropical ecosystems rely on figs as a keystone species. Fig wasps pollinate about 1,000 different varieties of figs.

Figs are unique because of how the flowers are contained within the immature fruit. To mate, lay eggs, and pollinate the small flowers, fig wasps enter through a tiny pore. Both are severe examples of obligatory symbiosis, in which the plant and the insect are entirely dependent on one another to survive.

### 2.7 Importance of pollinators

These small insects perform one of the most important ecosystem services on the planet, ensuring that both our culinary experiences and the world's environment flourish. Nearly 75% of the plant species cultivated for food, fibre, spices, beverages, condiments and pharmaceuticals are pollinated by animals (**Table 2**). The status of pollinator populations has huge economic impacts on agriculture.

| | | |
|---|---|---|
| 1. | Fruits, berries and nuts | Almonds, Apple, Apricot, Avocado, Blackberry, Blueberry, Cacao, Cashew, Cherry, Chestnut, Citrus, Coffee, Coconut, Cranberry, Date, Fig, Gooseberry, Grapes, Guava, Huckleberry, Kiwi, Litchi, Mango, Olive, Papaya, Peach, Pear, Plum, Pomegranate, Raspberry, Strawberry, Vanilla, Watermelon |
| 2. | Herbs and spices | Black Pepper, Cardamom, Chive, Clove, Coriander, Fennel, Lavender, Mustard, Nutmeg, Parsley, Pimento, Tea, White Pepper |
| 3. | Legumes | Beans, Cowpea, Lima Beans, Lupines, Mung Bean/Green or Golden Gram, Soybean |
| 4. | Seeds and grains | Alfalfa, Buckwheat, Canola, Flax, Oil Palm, Safflower, Sesame, Sunflower |
| 5. | Vegetables | Asparagus, Beet, Broccoli, Brussels Sprouts, Cantaloupes, Carrot, Cauliflower, Celeriac, Celery, Cucumber, Eggplant, Endive, Green Pepper, Leek, Lettuce, Okra, Onion, Parsnip, Pumpkin, Radish, Rutabaga, Squash, Tomato, Turnip, White Gourd |
| 6. | Others | Cotton, Kenaf |

**Table 2.**
*Common agricultural crops benefited by insect pollination [24].*

While some crops such as corn and wheat, are wind pollinated and some others like potatoes reproduce vegetatively, a whopping 35% of agricultural yield relies on animal pollinators [25]. Roubik published a comprehensive list of 1330 tropical crop species, including a list of viable breeding systems and pollinating taxa [24].

Williams examined the pollinator requirements for 264 crop species in Europe and found that 84 percent of them rely on animal pollination to some extent [26]. To put this in context, pollinators contribute over about $200 billion to the global economy [27].

The benefits of pollinators can easily be expanded to global biomes exceeding our gardens, kitchens, and dinner tables. With so many of the world's plants depending on pollinators for reproduction, these flower-loving friends are inadvertently supporting soil stabilisation, carbon sequestration and animal habitats. Sustaining healthy pollinator populations leads to supporting healthy ecosystems. The native pollinators not only provide a significant portion of the food and add to the economy, but they also play an important part in the natural ecosystem. The native pollinators help to keep the plant communities healthy and able to reproduce. They also support plants to provide cover and food for wildlife, to prevent erosion and keep waterways clean. The fruits and seeds produced by pollinated plants form an important part of the diet of birds and mammals. Many insects, including butterflies, use flowering plants as egg laying and nesting places.

## 2.8 Dependence on pollinators

The significance to a plant or the loss of its pollinators depends on whether the pollination relationship is facultative or obligate [28]. Some plants grow as a result of vegetative reproduction and are thus unaffected by the loss of pollinators. Others have vast seed banks or live a long time, so they may not be in immediate risk of extinction if their pollinator goes extinct. Most plants have several pollinators, and most pollinators pollinate multiple plant species, rather than a rigid one-pollinator-one-plant relationship. The composition of communities varies with environment, and what appears to be a specific relationship between a plant and a pollinator species may shift over time. Plants that are dioecious and self-incompatible, those with a solitary pollinator, and those that proliferate only by seeds are the most vulnerable to pollinator loss.

### 2.9 Decline of pollinators

Many pollinator habitats have been destroyed or disrupted as a result of human activities. Invasive plant species have fragmented and damaged many remaining habitat regions and such habitats become less suitable for pollinators and other wildlife. These habitat alterations may result decline in food sources, nesting and mating sites of native pollinators. Many pesticides have negative effects on pollinators and their habitats due to overuse and poor application. Herbicides diminish forage plant diversity by eliminating wildflowers, and some pesticides harm pollinators directly, particularly pollinating insects. Honeybees, for example, might outcompete indigenous pollinators for local nectar resources, putting them at greater risk of extinction. Pollinator populations have declined significantly as a result of habitat degradation and fragmentation. At least 185 pollinator species are designated as threatened or extinct by the International Union for Conservation of Nature (IUCN), and two bat species and 13 bird species are recognised as endangered in the United States.

### 2.10 Threats to pollinators

A number of threats to pollinators have been identified. These include habitat alteration, habitat fragmentation, introduction of alien pollinators and pesticide poisoning [28].

*Habitat alteration:* Agriculture, grazing, fragmentation of native landscapes and development of areas that once supported wild vegetation all are responsible for the loss of native food plants, rendezvous plants and nesting sites used by pollinators. Pollinators may depend on native plants because they are not always able to access food rewards from introduced flowers [29].

Many bees not only require large numbers of flowers to provide nectar and pollen, but also need a variety of flowering plants for their sustainability throughout the growing season. Oligolectic insects, such as some bees and butterfly larvae depend on specific plants for survival and persistence of their populations.

In addition to food requirements, pollinating organisms often have specific nesting requirements. Some bee species nest in cavities in the ground such as old rodent burrows, spaces under rocks, or holes excavated in sand or soft dirt. Many other types of bees nest in hollow twigs. As land is developed for human activity, the availability of twigs, rodent burrows and suitable nesting substrates typically decrease.

In the present scenario, large-scale monoculture of crops and intensive cropping practices reduce the amount of land available to support wild vegetation. With the increasing mechanisation of agriculture, the decrease in number and area of hedgerows and uncultivated patches reduced the number of native plants available as pollen and nectar sources [29, 30].

Gess and Gess determined that grazing livestock alters habitat sufficiently to affect pollinators [31]. They documented changes in availability of nesting sites, water resources, and vegetation that have direct negative effects on species diversity and population size of bees and wasps. Trampling of vegetation by livestock can directly destroy the nests of ground-nesting species and can compact the soil, constraining nest formation. In addition, the people who tend livestock in these areas of South Africa collect wood for fuel, thus reducing the availability of hollow twigs that provide nesting sites for some bee pollinators. Grazing also affects bees by decreasing water availability. Both ground-nesting and cavity-nesting bees must collect water for use in nest construction. Most bees cannot obtain water from livestock water tanks with steep sides, or even ponds without sloping edges, but need to stand at the edge of shallow water.

Tampering with the natural water supply to provision cattle or produce crops often modifies water availability for bees. Dramatic reductions in bee number and species diversity have been documented in areas of the Guana caste Province of Costa Rica that were deforested to support cattle [32, 33]. Vinson *et al.* observed a decrease in the number of pollen collecting bees in the destroyed forest areas [33].

*Habitat fragmentation:* Development can fragment natural habitats, isolating remnants of plant populations. Endangered plants often exist in "ecological traps" [34] surrounded by different habitats. Such plants may lack the genetic diversity that would allow them to colonise these different habitats. Small populations can also suffer from reduced pollen delivery or reduced quality of the pollen delivered. For example, Lamont *et al*. found that population fragmentation reduced fertility to zero in *Banksia goodii* (Proteaceae) which is a highly outcrossing species pollinated by birds (honeyeaters) and mammals (honey possums) [35]. Flowers in small populations either receive fewer visits from pollinators or receive pollen from sibling plants, which results in low seed production. Furthermore, small populations are sometimes bypassed by pollinators as some pollinators exhibit density dependent foraging behaviour, preferring large floral displays to isolated flowers.

Although habitat fragmentation is a problem, preserving large tracts of a particular vegetation type may not be enough to maintain pollinator populations. Janzen and colleagues censured euglossine bee populations in parks and reserves in Costa Rica and determined that even within the same park, different habitats vary dramatically in bee diversity [36]. Many of the bee species travel long distances to pollinate plants that do not occur within the habitats in which they were collected. This finding indicates that preservation of diverse patches within an area may be essential to maintain adequate pollinator populations.

*Introduction of alien pollinators:* It can have both beneficial and detrimental consequences and occurred both intentionally and accidentally. Honey bees have spread around the world, here they have become competitors with native bees, birds, and other pollinators, and bumblebees have also been introduced to islands and continents where they did not occur naturally. These introductions have sometimes benefited agriculture, but their consequences for native plants and animals can be deleterious.

Several studies have indicated that introduced honeybees decrease the foraging success of native pollinators by competing with them for resources [37–42]. Such example is provided by honeybees in Australia. Honeybees were introduced in Australia approximately 150 years ago, and so far they were considered beneficial to the native flora. However, Paron concluded in a recent study that honeybees may actually be harmful to the native flora as they may displace native pollinators, they may be ineffective at pollinating native flowers and they may interact in complex ways with native pollinators to reduce the amount and efficiency of pollen transfer [38].

*Pesticide poisoning:* Pesticide usage is another major problem for pollinators. Chemicals applied to crop plants and to rangelands can cause high bee mortality. In the United States, pesticide use has created local problems since late 1800s, but the problems increased drastically after World War II, when there was a substantial increase in the use of pesticides on crops, range lands, and forests [43]. Herbicides have also been applied extensively to control weeds in crops and along road sides, thus reducing the availability of the native wild plants that provide food for pollinators.

Foraging on pesticide-treated plants is a major source of bee mortality, yet honeybees are often expected to pollinate crops that have been treated with pesticides.

The susceptibility of bees to chemical poisoning is usually related to their surface area-volume ratio. Bumblebees are often more tolerant of pesticides than honeybees because of their smaller surface area-volume ratio and honeybees are in turn more tolerant than most small native bees. Chemical poisoning results in abnormal communication dances and mistakes in indicating distance and direction to food sources, in addition to direct mortality.

One source of pesticides that affects pollinators is the broad-spectrum insecticides used to control grasshoppers on rangelands in the South-Western United States. The rangelands are sprayed with these insecticides to save the grasses for cattle forage. The sprays kill many other insects in addition to grasshoppers, including local pollinators. The grasshopper-spraying campaigns overlap the flowering period of a number of endemic rangeland plants that grow among the grasses and many of these plants are listed as endangered or threatened [44]. Additionally, these campaigns also imbricate the period of emergence and active foraging of majority of the native bee species [45].

Another example of how pesticide application can affect plant reproductive success through its action on pollinators comes from the studies conducted in forests of New Brunswick, Canada [46]. These forest areas were sprayed with Matacil (aminocarb insecticide) to control spruce bud worm, *Choristoneura fumiferana* Clem referable to family Tortricidae of order Lepidoptera. The pesticide application adversely affected the native bees of families Andrenidae, Helictidae and Anthophoridae and syrphid fly populations. Several insects displayed convulsions followed by death. The native understory lilies namely *Maianthemum canadense* (Asparagaceae) and *Cornus canadensis* (Cornaceae) showed significant reductions in fecundity. Commercial blueberry fields in New Brunswick also suffered reduction in fruit set when adjacent forest lands were sprayed with Fenitrothion to control spruce budworm [47].

### 2.11 Attracting pollinators

An area must have sufficient food, shelter, water, and nesting grounds to lure local pollinators. To ensure that habitat demands are met, habitat management actions should be implemented. For instance, landowners can acquire, build, or plant extra nesting sites for bees and butterflies. Depending on the type of native pollinator targeted, various habitat management strategies are used.

Plant-appropriate vegetation: Planting gardens or meadows with a variety of native wildflowers, trees, grasses and shrubs is the easiest approach to attract local pollinators. Wildflowers and indigenous grasses will offer food such as nectar, pollen and larval host plants. For pollinators, trees and dense shrubs provide crucial shelter, nesting and overwintering places. Considering pollinator species have different preferences, planted areas should have diverse amounts of vegetation and areas of light, full shade and partial shade. Planting should take place in wind-protected areas.

Native plants should be chosen since these have evolved with local pollinators and are adapted to local soils and temperature. Native plants should make up at least 75% of a habitat's surface area. The cultivation of invasive species should not be avoided because such plants disrupt the ecosystem's natural structure and composition resulting in degrading pollinator and other wildlife habitat. The area of mowed lawn should be restricted in favour of native wildflowers, shrubs, and grasses. The existing lawns should be mowed less frequently to allow plants to offer pollinator habitat. Annuals should be avoided in favour of perennials. Perennials are often higher in nectar content and provide a more reliable food source than annuals because they bloom year after year. Plants that reproduce in "doubles," such as marigolds and roses, should be avoided because such plants are designed for ornamentation rather than pollen and nectar availability. The species of wildflowers should be grown in a clump to attract more pollinators and not grown individually. Throughout the growing season, nectar and pollen flowers should be available. The variation in flower shape and colour will deliver nectar and pollen to a variety of pollinators. Bell, tube, or trumpet-shaped flowers, as well as those with clusters of tubular florets, are favourites of birds and butterflies, especially when surrounded with a flat surface for perching. They favour flowers that are brilliantly coloured such as oranges, yellows and reds. Yellow, blue, and purple flowers are most appealing to bees. The flowers that bloom at night attract moths and bats.

Use pesticides carefully: Pesticides, the chemical toxins, do not distinguish between beneficial and harmful insects. As an insecticide is used to kill a crop-eating insect, it may also harm important natural pollinators. Pesticide treatment has the potential to harm or kill all pollinator species, as well as to effect other wildlife. Pollinators can be poisoned by such chemicals through contaminated food or directly from the contaminated surfaces of florets, leaves, soil, or other things when they come in contact with them. To sustain the whole spectrum of native pollinators, usage of such chemicals should be restricted or kept to a bare minimum. To address pest infestations, landowners should use non-chemical or organic methods.

Provide water: The pollinator species require water to survive. Bees and butterflies should be attracted to a source of pesticide-free water mud and other beneficial insects drawn to a birdbath, fountain, tiny pond, or mud puddle. For butterflies and bees, a moist salt lick can be made. A damp patch on the earth can be created by using a dripping hose, drip irrigation line, or birdbath and additionally, a small amount of sea salt or wood ashes can be mixed to meet the mineral needs of butterflies and bees.

## 3. Conclusions

Insects, being diverse and dominant, are the key component of a healthy ecosystem. Humans determine whether an insect is beneficial, benign or pestiferous. Majority of them are beneficial to humans either directly or indirectly as food, pollinators, pollution indicators, scavengers, for production of useful products etc. The insects represent their dominance as pollinator. Bees and flies visit more than 90% of the world's major plant types, while the other species visit fewer than 6% of the crop varieties. The effectiveness of pollinators varies according to factors such as their abundance; their ability to reach individual plants of the same species and to collect, transfer and deposit the pollen to the appropriate plant organ. Insect pollinators are in decline which is tentative, considering the lack of comprehensive data [48], but it is still a matter of concern. Losses in diversity and abundance are particularly strong under intensive agricultural management [49, 50]. Despite their significance, pollinators are declining and often overlooked in terms of their contributions to healthy ecosystems. No pollinators would mean no seeds or fruits and therefore the collapse of agriculture. No plant reproduction in the wild means that many plants will become locally extinct. Human activities have destroyed and fragmented native pollinator habitats. This diversity needs protection by integrating conservation measures with sustainable agricultural practices, which may raise crop yields and protect both wild and managed species of bees and other pollinators.

A range of conservation measures in intensively-farmed regions can help to maintain diversity, by preserving the resources that pollinators need. Some of the measures are at farm-level such as planting flower strips among crops, reintroduction of hedges and planting trees while as others are implemented at landscape-level such as the conservation of natural and semi-natural habitats in agricultural landscapes. There is no "one size fits all" approach to conserve all species, due to their varying preferences for different food sources and nesting sites. Reversing the decline in pollinators is the key to feed mouths in future and must be seriously given a thought and action plan.

## Conflict of interest

The authors declare no conflict of interest.

## Author details

Navkiran Kaur* and Amritpal Singh Kaleka
Punjabi University, Patiala, Punjab, India

*Address all correspondence to: navkiran.dandiwal@gmail.com

## References

[1] Reverté S, Retana J, Gómez JM, and Bosch J. Pollinators show flower colour preferences but flowers with similar colours do not attract similar pollinators. Annals of Botany. 2016; 118: 249-257. DOI: 10.1093/aob/mcw103

[2] Wasserthal LT. The pollinators of the Malagasy star orchids *Angraecum sesquipedale*, A. sororium and A. compactum and the evolution of extremely long spurs by pollinator shift. Botanica Acta. 1997; 110: 343-359.

[3] Ollerton J, Rachael W, Sam T. How many flowering plants are pollinated by animals? Oikos. 2011. DOI:10.1111/j.1600-0706.2010.18644.x.

[4] Kearns CA. Endangered mutualisms: the conservation of plant–pollinator interactions. Annual Review of Ecology and Systematics. 1998; 29: 83-112.

[5] Tepedino V. The importance of bees and other insect pollinators in maintaining floral species composition. Pages 39-150 in great basin naturalist memoirs nr 3: the endangered species a symposium; 7-8 Dec: 1979. Provo (LT): Brigham Young University.

[6] Levin DA. The origin of reproductive isolating mechanisms in flowering plants. Taxon. 1971; 20: 91-113.

[7] Michener CD. 2007. The Bees of the World. 2nd edn. 2007. pp. 992. Johns Hopkins University Press, Baltimore, Mary-land, USA.

[8] Kleijn D, Winfree R, Bartomeus I, Carvalheiro LG, Henry M, Isaacs R, Klein AM, Kremen C, M'Gonigle LK, Rader R, Ricketts TH, Williams NM, Lee Adamson N, Ascher JS, Baldi A, Batary P, Benjamin F, Biesmeijer JC, Blitzer EJ, Bommarco R, Brand MR, Bretagnolle V, Button L, Cariveau DP, Chifflet R, Colville JF, Danforth BN, Elle E., Garratt MPD, Herzog F, Holzschuh A., Howlett BG., Jauker F., Jha, S., Knop, E., Krewenka, K.M., Le Feon, V., Mandelik Y., May, E.A., Park, M.G., Pisanty, G., Reemer, M., Riedinger, V., Rollin, O., Rundlof M., Sardinas H.S., Scheper, J., Sciligo, A.R., Smith, HG., Steffan-Dewenter, I., Thorp, R., Tscharntke, T., Verhulst, J., Viana, B.F., Vaissiere, B.E., Veldtman, R., Westphal, C. and Potts SG. Delivery of crop pollination services is an insufficient argument for wild pollinator conservation. Nat Commun. 2015; 6: 7414. https://doi.org/10.1038/ncomms8414

[9] Neff, JL., Simpson, BB. Bees, pollination and plant diversity. In: Gauld, I.D., LaSalle, J. (eds.) Hymenoptera and Biodiversity. 1993. CAB International, Wallingford

[10] Klein, A.M., Vaissiere, B.E., Cane, J.H., Steffan-Dewenter, I., Cunningham, S.A., Kremen, C., Tscharntke, T. Importance of pollinators in changing landscapes for world crops. Proceedings of the Royal Society of London series B. 2007; 274: 303-313.

[11] Larson, B. M. H., P. G. Kevan and D. W. Inouye. Flies and flowers: I. The taxonomic diversity of anthophiles and pollinators. Canadian Entomologist. 2001; 133(4): 439-465.

[12] Elberling, H. and Olesen, JM., 1999: The structure of a high latitude plant-flower visitor system: the dominance of flies. Ecography. 1999; 22: 314-323.

[13] Scoble MJ. The Lepidoptera: form, function, and diversity, 2$^{nd}$ ed. 1995. Oxford University Press, Oxford

[14] Kevan PG, Baker HG. Insects as flower visitors and pollinators. Annual Review of Entomology. 1983; 28: 407-453.

[15] Proctor, M., Yeo, P. and Lack, A. The Natural History of Pollination. 1996. Harper Collins, London.

[16] Willmer, P. Pollination and Floral Ecology. 2011. Princeton University Press, NJ, USA Nagamitsu and Inoue, 1997

[17] Nagamitsu, T. and Inoue, T. Aggressive foraging of social bees as a mechanism of floral resource partitioning in an Asian tropical rainforest. Oecologia, 1997;110: 432-439.

[18] Wester, P., Stanway, R. and Pauw, A. Mice pollinate the Pagoda Lily, *Whiteheadia bifolia* (Hyacinthaceae) - First field observations with photographic documentation of rodent pollination in South Africa. South African Journal of Botany, 2009; 75: 713-719.

[19] Yumoto, T., Momose, K. and Nagamasu, H. A new pollination syndrome – squirrel pollination in a tropical rainforest in Lambir Hills National Park, Sarawak, Malaysia. Tropics, 1999; 9: 133-137.

[20] Olesen, JM, Valido, A. Lizards as pollinators and seed dispersers: an island phenomenon. Trends in Ecology and Evolution, 2003; 18: 177-181.

[21] Hansen DM, Beer K, and Muller CB. Mauritian coloured nectar no longer a mystery: a visual signal for lizard pollinators. Biology Letters. 2006; 2: 165-168.

[22] Ortega-Olivencia, A., Rodríguez-Riaño, T., Pérez-Bote, J.L., López, J., Mayo, C., Valtueña, F.J. and Navarro-Pérez, M. Insects, birds and lizards as pollinators of the largest-flowered Scrophularia of Europe and Macaronesia. Annals of Botany, 2012; 109: 153-167.

[23] Sarma, K., Tandon, R., Shivanna, KR. and Mohan Ram, HR. Snail pollination in Volvulopsis nummularium. Current Science, 2007; 93: 826-831.

[24] Roubik, DW. Pollination of cultivated plants in the tropics. 1995. Food and Agriculture Organization of the United Nations, Rome.

[25] Klein, A.M., B.E. Vaissiere, J.H. Cane, I. Steffan-Dewenter, S. Cunningham, C. Kremen, and T. Tscharntke. Importance of pollinators in changing landscapes for world crops. Proceedings of the Royal Society B: Biological Sciences, 2007; 274(1608): 303-313.

[26] Williams, IH. The dependence of crop production within the European Union on pollination by honeybees. Agricultural Zoology Reviews. 1994; 6: 229-257.

[27] Gallai N., Salles, JM., Settele, J. and Vaissiere, BE. Economic Valuation of the Vulnerability of World Agriculture Confronted with Pollinator Decline; Ecological Economics, 2009; 68(3): 810-821.

[28] Bond, WJ. Do mutualisms matter? Assessing the impact of pollinator and fdisperser disruption on plant extinction. Philosophical Transactions of the Royal Society of London Series B – Biological Sciences. 1994; 344:83-90.

[29] O'Toolc C. Diversity of native hees and agroecosystems. 1993. Pages 169-196 in LaSalle J, Gauld ID, eds. Hymenoplera and biodiversity. Oxon (UK): C.A.B. International.

[30] Williams PH. 1986. Environmental change and the distributions of British bumble bees (*Bombus* LHr.). Bee World. 1986; 67: 50-61.

[31] Gess FW, Gess SK. Effects of increasing land utilization on species representation and diversity of aculeate wasps and bees in the semi-arid areas of southern Africa, 1993. pp. 83-113. In Hymenoptera and Biodiversity. (Edited by J. LaSalle and I. D. Gauld). CAB International, Wallingford, UK.

[32] Janzm DH. The deflowering of Central America. Natural History. 1974; 83: 49-53.

[33] Vinson SB., Frankie GW, Barthell J. Threats to the diversity of solitary bees in a neotropical dry forest in Central America. 1993. Pages 53-82 in LaSalle J, Gauld ID, eds. Hymenoptera and biodiversity. Oxon (UK): C.A.B. International.

[34] Stebbins GL. Rare species as examples of plant evolution. 1979. Pages 113-118 in Great Basin naturalist memoirs nr 3: the endangered species; 7-8 Dec 1979. Provo (UT): Brigham Young University.

[35] Lamont BB, Klinkhamer PGl and Witkowski ETF. Population fragmentation may reduce fertility to zero in *Banksia* goodiia demonstration of the Allee effect. Oecologla. 1993; 94: 446-450.

[36] Janzen DH, Devries P, Higgins MI and Kimsey LS. Seasonal and site variation in Costa Rican euglossine bees at chemical baits, in lowland deciduous and evergreen forests. Ecology. 1982; 63: 66-74.

[37] Gimhurg HS. Foraging ecology of bees in an old field. Ecology. 1983; 64: 165-175.

[38] Paron DC. Honeybees in the Australian environment. BioScience. 1993; 43: 95-101.

[39] Pyke GH, Balzer L. The effects of the introduced honeybee (Apis mellifera) on Australian native bees. 1985. Ocasional paper nr 7. Sydney (Australia): New South Wales National Parks Wildlife Service.

[40] Roubik DW, Moreno JE, Vergara C, Wittmann D. Sporadic food competition with the African honey bee: projected impact on neotropical social bees. Journal of Tropical Ecology. 1986; 2: 97-111.

[41] Schaffer WM, Jensen DB, Hobbs DE, Gurevitch J, Todd JR., Schaffer. MV. Competition, foraging energctics and the cost of sociality in three species of bees. Ecology. 1979; 60: 976-987.

[42] Sugden EA, Pyke GH. Effects of honey bees on colonies of *Exoneura asimillima,* an Australian native bee. Australian Journal of Ecology. 1991; 16: 171-181.

[43] Johansen CA. Pesticides and pollinators. Annual Review of Entomology. 1977; 22:177-192.

[44] Bowlin, W.R., V.J. Tepedino, T.L. Griswold. The reproductive biology of *Eriogonum pelinophilum* (Polygonaceae). Proc. Southwestern Rare and Endangered Plant Conf., Santa Fe, 1993. pp. 296-302.

[45] Peach, ML, Tepedino, VJ, Alston, DG, Griswold, TL. Insecticide treatments for rangeland grasshoppers: potential effects on the reproduction of *Pediocactus sileri* (Englem.) Benson (Cactaceae).1993. In: R. Sivinski; Lightfoot, K., eds. Proceedings of the southwestern rare and endangered plant conference. Misc. Pub. 2. Santa Fe, NM: New Mexico Forestry and Resources Conservation Division: 309-319.

[46] Thomson JD, Plowright RC, Thaler CR. Matacil insecticide spraying, pollinator mortality, and plant fecundity in New Brunswick forests. Canadian Journal of Botany, 1985; 63: 2056-2061.

[47] Kevan PG. Forest application of the insecticide Fenitrothion and its effect on wild bee pollinators *(Hymenoptera:Apoidea)* of lowbush blueberries *(Vaccimum* spp.) in Southern. New Brunswick, Canada. Biological Conservation. 1975; 7: 101-309.

[48] LeBuhn G, Droege S, Connor EF, Gemmill-Herren B, Potts SG,

Minckley RL, Griswold T, Jean R, Kula E, Roubik DW, Cane J, Wright KW, Frankie G, Parker F. Detecting Insect Pollinator Declines on Regional and Global Scales. Conservation Biology: The Journal of the Society for Conservation Biology, 2012. December, 1-8. doi:10.1111/j.1523-1739. 2012.01962.x.

[49] Biesmeijer, JC, Roberts, SPM., Reemer, M., Ohlemuller, R., Edwards, M., Peeters, T., Schaffers, AP., Potts, SG., Kleukers, R., Thomas, CD., Settele, J. and Kunin, WE. Parallel declines in pollinators and insect-pollinated plants in Britain and the Netherlands. Science. 2006; 313: 351-354.

[50] Le Feon V, Schermann-Legionnet A, Delettre Y, Aviron S, Billeter R, Bugter R, Hendrickx F, Burel F. Intensification of agriculture, landscape composition and wild bee communities: A large scale study in four European countries. Agriculture, Ecosystems and Environment. 2010; 137: 143-150.

Section 2

# Possible Mitigating Measures of Insect Decline

Chapter 6

# Botanical Insecticides Are a Non-Toxic Alternative to Conventional Pesticides in the Control of Insects and Pests

*Nazeer Ahmed, Mukhtar Alam, Muhammad Saeed, Hidayat Ullah, Toheed Iqbal, Khalid Awadh Al-Mutairi, Kiran Shahjeer, Rafi Ullah, Saeed Ahmed, Nibal Abd Aleem Hassan Ahmed, Hanem Fathy Khater and Muhammad Salman*

## Abstract

Insect control for crops is one of the most critical global concerns. Pest management is an economic and ecological problem worldwide due to the human and environmental risks raised by most synthetic pesticide products. Botanical insecticides have resurfaced in popularity due to their low cost and low environmental impact, rather than their negative effects on human health. Botanical insecticides destroy only the insects they are meant to kill, leaving no residue on food or in the environment. Botanicals have long been used to combat pests. The compounds have many environmental advantages. However, as opposed to other bio-control pests and pathogens, their use was minimal during the twentieth century. In developing countries, botanical insecticides are well adapted for use in organic food production. Nonetheless, they may play a far bigger role in developed countries' food production and post-harvest food protection. Consequently, the current chapter briefly addresses botanicals with active ingredients with insecticidal, antifeedant, or repellent properties.

**Keywords:** insect, crop protection, active constituents, insecticides, natural products, action mechanism

## 1. Introduction

Insects are the world's most abundant animal species, and they can be found in any ecosystem. Pest insects account for fewer than 0.5 percent of all insect species, and just a few are dangerous to humans. Certain insects can be dangerous to entire countries or groups of countries [1]. Crops are continuously at risk of being infested or infected. Since pesticides are cheap and easily applied, farmers typically use fast pest control measures like synthetics to protect their animals and crops from infestation. Synthetic pesticides can tend to select more pesticide-tolerant ones in the population, but it does lead to developing pesticide-resistant pests. To oversimplify and misuse synthetic pesticides in agriculture can damage human health and the environment, even damaging biodiversity Research suggests that constant consumption of synthetic pesticides can cause human illnesses and diseases [2–4]. Furthermore, most synthetic pesticides are not biodegradable, causing soil and groundwater contamination and ozone depletion in the atmosphere. The negative consequences of misuse and overuse of synthetic pesticide have prompted alternative pest control solutions [5–7].

Plants containing bioactive chemicals have been shown to effectively treat a variety of crop pests and human illnesses [8, 9]. Plants like pyrethrum (*Tanacetum cinerariifolium*) and anemone (*Anemone Brizo*) have been shown to have pesticide and malaria-control properties in their insect repellent abilities. Human management of plant problems was practiced, and pesticides were gradually phased out by humans, rather than being replaced by technology and newer, more toxic, but more effective pesticides. They had great success in combating serious plant diseases such as rust and blight, where they are more effective and less toxic, and they became very popular [10]. As a result, natural plant-based products were gradually phased out until recently, when synthetic pesticides threatened human health and the environment [11]. Today, people want food grown with pesticide-free and naturally derived treatments. At the same time, detection of toxic pesticide residues in food and a heightened interest in food safety has prompted agriculture-organic bans of certain chemicals [12, 13].

Continued usage of synthetic insecticides has caused environmental damage, health problems, and loss of species diversity, contamination and biodiversity problems, and an increase in exposure to danger to hazards [14]. Synthetic pesticides have harmed farmers in the export trade, especially in the horticultural sector [15]. Both farmers and exporters in developing countries have lost market and profits if banned pesticides are detected above-defined tolerable level. Alphadime® (alpha-cypermethrin + dimethoate) and Demeton®, for example, are no longer allowed to be used on fresh produce exported to other countries [16, 17].

All the aspects that contribute to the value of botanical pesticides are efficacy, biodegradability, various modes of action, low toxicity, and the accessibility of the source materials. Pre- and pre-harvest times are frequently small [18]. In organic agriculture, where organic food attracts higher costs, botanical pesticides are extensively used [19]. As a result, botanical insecticides are becoming popular since they are safe in crops cultivated for human consumption, and customers willing to pay an organically cultivated premium are more demanded [20]. Many investigations have been carried out with known and still to be utilized species of plants having pesticide characteristics [21, 22]. The commercially available botanical pesticides are examples of pyrethrum (*Tanacetum cinerariifolium*), neem (*Azadirachta indica*), sabadilla, tobacco (*Nicotiana tabacum*) and ryania (*Ryania speciosa*) [23]. In post-harvest pest control, farmers have traditionally utilized plant protection agents, particularly for grain conservation, while they were storing.

The derivatives of plant products that repel, inhibit, or destroy pest are botanical pesticides [24]. Many studies have concentrated on managing pest populations using different botanical pesticides to control insects [25–29]. Plants with pesticide properties can deal with bacteria, fungi, and nematodes; likewise, toxins affect pests. This chapter features data on the chemical composition of botanicals, their pest-control mechanisms, the problems of their use, and the need for them.

## 2. Background history

Plants have been used as pesticides since humans discovered that some plants defend themselves better than others. Before using any other pesticides, people used botanicals to combat pests. They are recorded in hieroglyphic, Chinese, Roman, and Greek antiquities. In India where the neem tree of the Veda, a collection of handwritten archeological Sanskrit written at least 4000 years old, has been mentioned Neem (*Azadirachta indica* Juss.; Meliaceae). Plant compositions for controlling insect pests were mentioned in various writings of the 18th century. In the late 19th century, poisonous plants or minerals were often applied such as oils, tars, sulfocalcic sprinklers, hot water and other technologies [30]. Plant extracts were created as a result of combining empirical and scientific findings.

The first pesticides were made with readily available botanicals and allelochemicals. Since pest insects are easier to identify, they were targeted rather than pathogens. Biopesticides of plant origin have been studied in many recent books and chapters [31, 32].

Plant development as pesticides has two sources of development: First, there are historical and existing uses of plants and their plant constituents in cattle and crop protection methods; and second, the analysis of plant extracts for active ingredients and plant protection. Nicotine activity obtained from tobacco *Nicotiana tabacum, Derris elliptica,* and rotenone from *Leguminosae Lonchocarpus* fall into this group. (ii) systematic sampling of plant families obtained in searching campaigns to detect active molecules, accompanied by biological tests. Such prospecting was done in the 1940s with the help of Rutgers University and Merck, and the outcome was Ryanodine, an alkaloid derived from *Ryania* sp. that was first sold in the United States in 1945 [31].

Four major compounds were widely used before WWII: Alkaloids and *nicotines, rotenone, pyrethrins,* and vegetable oil. *Nicotines* and alkaloids The usage of these compounds waned due to their toxicity to nicotine organisms or molecular instability (pyrethrum), while chemically synthesized pesticides were marketed during WWII (organochlorides, organophosphates, and carbamates). Their management was cheaper and easier. Until the 1960s, this condition persisted [31].

However, a resurgent interest in botanicals was shown by several demonstrations that the widespread use of chemical pesticides can adversely impact non-target creatures and environmental hazards. Even though a great effort was made in the second half of the twentieth century to search for and produce newly synthesized pesticides, research was conducted on plant-based Biopesticides to increase their stability or discover novel compounds and molecules. An excellent illustration of this is the syntheses is of pyrethroids, pyrethrum-derived synthetic compounds, and neem (*Meliaceae*) in the 20th century.

## 3. Botanical pesticides sources

Some botanical pesticides were be obtained from plants extracts essential oils, or combinations. Certain plants are known to be used as botanicals. Rhizomes, bark, leaves, nuts, cloves, fruits, and stems are ingredients. In this context, the application of the plant component would rely on which bioactive compounds are utilized and their levels of abundance within the target cells. Botanical insecticides are manily found in the following plant families: *Myristicaceae*, *Rutaceae*, *Caesalpinaceae*, *Apiaceae*, *Caesalpinaceae*, *Sapotaceae*, *Cupressaceae*, *Piperaceae*, *Solanaceae*, and *Zingiberaceae* [33–35]. Dried and pulverized plant parts are extracted using solvents that promote extraction. After the extraction, the potency is distilled, standardized, and tested in a laboratory or field. Other examples of viable and profitable botanical pesticides have included the neem herb *azadirachtin* (Azadirachta) and the insect repellent pyrethrum (*Tanacetum cinerariifolium*). Many other plants have pesticide properties like Garlic (*allium sativum*), Turmeric (*Curcuma longa*), Rosemary (*Rosmarinus officinalis*), Ginger (*Zingiber officinale*), peppermint, (Mentha piperita), and Thyme (*Thymus vulgaris*) [36, 37].

## 4. Factors that affect botanical pesticides

1. Supply of raw materials.

2. Botanical extract standardization containing the dynamic combination of active ingredients.

3. Types of solvents, plant organisms, and plant parts.

4. Quick decomposition and Environmental factors.

5. Market prospects for botanical pesticides.

6. State registration.

Some factors that influence the usage of synthetic botanical pesticides include the pesticide's composition, the active ingredient, method, time, and the quantities used in the mixtures of pesticides, climatic conditions and the time of year of application [38].

Thus, an investigation must also consider possible environmental exposure, indicators of health, and other aspects of risk assessment such as an individual's residency and work background, clinical history, and the prevalence, in the area in which populations are examined of pesticides analyzed in drinking water, land, atmosphere and fresh and processed food. The length of time spent each day, the number of years spent conducting the activity, the type of exposure, the use of protecting facilities, and their geographical closeness to agricultural fields can increase exposure [39].

## 5. Botanical insecticides made from plants are used in agriculture

This class of plants is of prime importance as botanical pesticides, herbs, or ornamental plants, can be found in the environment, and a lot of them serve several purposes such as medicines, foodstuffs, accessories, and livestock. They are widely available, and thus very economical, and thus can be easily adopted into agricultural practices. Neem, pyrethrum, and several other non-target species, commercially sold pesticides, are the least harmful, such as insects and fish to none target organisms. They are healthy for both human use and the climate.

The relationship between plant-derived pest-control products and pests is based on a biochemical process, which will decrease the likelihood of resistance. Essential oils and essential extracts have a derivative focus on target-specific properties, which help protect bees and other non-target beneficial species from a plant-based risk. Has no or little allelopathic impact on botanical crops Its effectiveness depends on the plant species, whether the extract is used dry or liquid, solve concentrations, and extraction methods. They have a variety of modes of action including insect resistance, population control, toxicity, and crop modification to meet a variety of different pests' requirements. They interact with behavioral activity, metabolic processes, anatomy, biochemical activities, and certain physiological functions. For example, the terpenoids interfere with phenomenology on moth phenology cells (**Figure 1**) [40–49].

Some scientists have critically examined the acceptance, adoption, and use of botanical pesticides. There must be enough knowledge and proof of the chemistry and effectiveness of botanical pesticides before they can be approved for general use. These provide details on the composition, degradation, durability, and toxicity of the substances [50].

Food safety is enhanced due to the integration of botanicals in agricultural systems, particularly in greenhouses, production; crop productivity is improved through increased and greater market accessibility thanks to that, along with higher prices due to lower pest densities; and guaranteed market access. A certain subset of the consumer population is willing to pay more for organically grown foods, and this opens the door for the botanical pesticides that are profitable for the farmers to expand their market share of that population. **Figure 1** shows the various pathways that can be followed when considering both synthetic and botanicals. Synthetic pesticides contribute to agriculture have the benefit of reducing crop damage and cutting the number of money farmers have to spend on pesticides and increases in sales and profits on their produce. At the same time, these methods must be used judiciously and by skilled staff should be implemented. IPM systems that incorporate botanical pesticides will eliminate the overuse of synthetic pesticides instead of the more common practice of using either of the two. For these reasons, small farmers and family farmers need to take proper precautions and ensure both human and environmental protection; [51, 52].

In this chapter, we'll look at using botanical products to control insects in crop production. From a chemical standpoint, we offer a summary of botanical insecticides and classify their effects on insects.

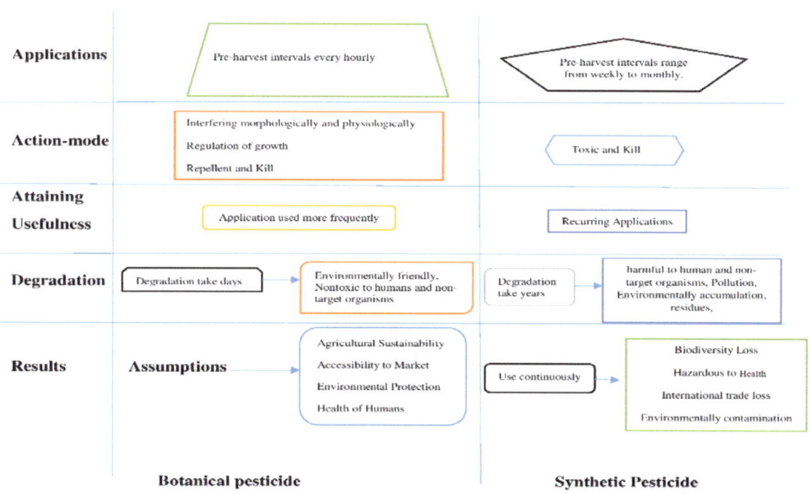


**Figure 1.**
*Differences between botanical and synthetic pesticides with respect to mode of action, use, persistence and effect on ecosystem.*

## 6. Botanical insecticides types

### 6.1 Fatty acids and esters

The single application of allyl cinnamate can result in highly toxic effects in the S. littoral larval stages of the cabbage whitefly and onion maggot. Ethyl (E, Z, E)-2-decent (Zeder's) was confirmed to be an effective insecticide against the Cimex, while Schmidt et al. [53] were unable to obtain an example for testing. Studies on fat-metered homogenate suggested that fat methyl esters (derived from *Solanum chlamygynidense*) have larvicidal properties for the cinque feed vector, *Culex. quinquefasciatum* [54]. Studies show that saturated and polyunsaturated acids (particularly C8, C9, and C10) work against houseflies, horn flies, and stable flies, respectively a fatty acid mixture (C8910) has shown to be both toxic and refractive to an insecticide-resistant Anopheles mosquito strain. In the literature by Youssef et al. [55], it was found that the larvicidal activity of linoleic acid was active against *S. littorals*, and the larval weight was reduced.

### 6.2 Glycosides

In general, plants use cyanogenic glycosides in defense against their herbivores, although some species have been observed to use them for purposes of protection as well as for damage by certain pests. Velu et al. [54], discovered that the digitized glycoside (purple root), sourced from *Digitalis purpura*, *calotropis procera*, was effective against both larval and adult stages of the camel tick ticks, as well as against *Azadirta andneemidos* genuses, while "kinds, in combination with *hyaliqueinul* bearing Neem oil and *Proxeebrin acedra* on a bioassay, showed digitoxin from *purpurean digitalids* could hold against various species of camel, while proven in addition to all lar and adult stages of *Hyommadromesis rajene* had the proper concentration. Additionally, it discovered that Viscin-2 and Vtsin, which serve as growth-inhibiting photo plastic herbicides for insects, also prevent the larva of the cotton aphid species from gaining weight. Have acridglycides (from Bothidae and Mucroneidae), they do not possess the protein insect binding of gypsophilla (*L. dispar*, *N. coenia*, and, to be more precise, juneids (*Lymriidae* and *Mucranidae*) do). Since cyanogenic glycosides are found in cassava or other plants, it's also believed that they are components of these plants' plant defense mechanisms. Like most stored-product insecticides, they are effective against: they are effective against both pests and infestations. Cyanohydroarilase has pesticide properties to the lepidozinans (particularly in indoor areas), which means that it is useful as alternative pest control and can be applied to the soil as fumarate [55]. This was discovered by (in this study) on species from the genus of ants of Cassia, which possess a *malathion peroxide* (antimalaria) and are frequently used in malarious regions as antimalarial/insecticidal activity. The larvae found in *A. gigas*, chinch Bugs from Glycinequa have larvicide activity against chinch and malaria visas vectors. The effectiveness of juvenile hormone treatments in pest control is outstanding in recent experiments [56].

### 6.3 Flavonoids

Flavonoids have the potential to be effective in pest-control measures. Flavonoids are important in protecting plants from insect pests and herbivores that feed on plants. Plants are protected from insect pests by flavonoids and isoflavonoids, influencing their behavior, growth, and development. *Pinus banksiana's rutin* and quercetin-3-glucoside inhibit the growth of *Lymantria. dispar* and increase its mortality. Tobacco armament (*Spodoptera litura*) death rates of peanuts enriched in quercetin and rutin glycosides increased. In *Nilaparvata lugens* and herbivores, three flavone glucosides present in rice impede insect digestion [55]. Insecticide activity against *Callosobruchus. maculatus* grubs, flavonoid glycosides derived from *Tephrosia purpuria* were exhibited. Two further forms of flavonoids protecting plants against insects are isoflavonoids and proanthocyanidins. For instance, *narengine procyanidine* suppresses *Aphis craccivora* and herbivores' growth [55]. Quercetin/azadirachtin insecticide can be a safe, efficient insecticide that increases the functioning and non-toxicity of *Euphaedra orientalis* [57]. It is also less environmentally damaging because it is quickly biodegradable. *Acyrthosiphone pisum* was identified by Goawska, Sprawka, Ukasik and Goawski [58], as polyphene-naringenin flavonol (*flavanone naringenin* and *flavonol quercetin*) as a pesticide against Pea aphid. (Aphididae, Hemiptera).). *Tagetes erecta* and *Tagetes patula* contain toxic plant chemicals (flavonoids) that can support their usage in the form of natural insecticides. Quercetin, Kaempferol and RCO, tricin, apigenin + RCO, apigenin and apigenin are efficient insecticides.

### 6.4 Alkaloid

Alkaloids are vital to insect control as they are among the most effective natural insecticides in nature. The authors concluded that pyridine alkaloids from castor bean proved effective against the malaria-carrier mosquito species *Anopheles gambiae*. The oil extracted from the leaves of *Ruta chaloderma* powder and quinone herbals had larvicidal and antifedi parasiticidal activity against caterpillars such as the larvae of the coastal *helio thopygea* butterfly. Antifeedant and larvicidal effects were found in the pergularia root alkaloids extract, that regular antifeedant and larvicide did Praline and piperidine alkaloids have activity against mosquito larvae *Arachis hypogaea* alkaloid has a larvicidal function [58, 59].

### 6.5 Nicotine

Nicotine, the addictive component of tobacco, is a tranquilizer in tobacco plants (*Nicotiana Tobacco*) and other Nicotiana species. Heavy doses because of respiratory paralysis are also exceedingly toxic. Nicotine is a ganglion cholinergic agonist with a wide spectrum of pharmacological effects mediated via autonomy, supranational medulla, neuromuscular crossover, and brain bonding to receptors [60].

### 6.6 Essential oils

Regnault-Roger and Philogne [61] state that natural chemical pesticides are plant extracts that are excellent alternatives to biological or synthetic pesticides. Additionally, chemical pesticides are difficult to use because of insect resistance to synthetic compounds, which has resulted in billions of dollars of food production losses annually. In addition, the United States Food and Drug Administration (FDA) accepted that botanical pesticides (essential oils), which are protected from non-target and cross-and multi-resistant to insects, are more likely to cause ozone depletion, neurotoxicity, carcinogenicity, teratogenicity, and mutagens [61].

The rising use of plant-based insecticides by organic growers has increased aromatics in essential oils extraction due to the rise in plant-based products and health-conscious consumers. These ingredients are used to kill and repel insects [60–62]. According to some researchers, essential oils are effective against bedbugs, ants, moths, and particularly the predatory, voracious, and especially larvae of the Gypsy moth, some types of insects. One observes that Peppermint oil is effective against Ants, Flies, Nips, and *Varroa most* stumptica; additionally, proves that it is effective against Both *Callospora*, *Tribrix*, and *Varrota* powdery locust. *Trichosomyia of uremia* larvae has Larvicide effective against *Aedes aegypti* mosquitoes and *C. Quinquefasciatus* [58, 61–64].

Nepetalactone is a very good active element for the repellent of mustaches, bees, and other flying insects in Catnips (*Nepeta cataria*). In repelling mosquitoes, it is more effective than DEET. The *Aedes aegypti* mosquito, which distributes the yellow fever virus, is highly effective. In contrast, *Zingiber* Official Rhizome and *Piper Cubeba* berries oil exhibited insecticide and anti-favoring activities in *Tribolium castaneum* and *Sitophilus oryzae*. *Tagetes* species oil exhibits an anti-insect effect against *Ceratitis capitata* and *Triatoma infestans*. *Melaleuca alternifolia's* fumigant toxicity to *Sitophilus spp*. Healthy cockroach repellents are rosemary, oregano, yarrow, eucalyptus and mint oil. *Supella longipalpa* is an oregano oil-killing parasite. It detected that insecticidal in larvae from the pine procession moth, *Thaumetopoea pityocampa*. *Laurus nobilis* essential oil has also been discovered toxic against *rhyzopertha dominica*, and *T. castaneum*. *Lavandula hybrida*, *Rosmarinus bureinalis* and *Eucalyptus globulus* have killed the adults of *Acanthoscelid obtectos*. *Tagetes minuta* essential oil has also been Acaricidal and repellant for *cochliomyia macellaria*. Basil oils contain eugenol, a potent anti-mosquito, and linalool, a harmful substance to *Bruchid zabrotes fasciatus* and other pests. *Lasioderma serricorn* repels *zingiber zerumbet's* essential oil. *Juniperus procera* essential oil has been proven to help repel *Anopheles arabiensis* malaria mosquito. All of the instances include terpinene-4-ol, 1,8-cineol, verbenone, and field horn. Anti-piling insects, insecticides and mosquito bites in adults were prevented by Eukalyptus oil, *Aedes aegypti* larvae have poisonous substances of *Eucalyptus globules*. Burning *Eucalyptus citriodoric* sheets are used as a mosquito repellant in Africa. Moreover, the CDC (Centers for Disease Control and Prevention, USA) advocated utilizing the West Nile virus to protect it from neurological disease and even from death and transmitted by mosquitos using the lemon *eucalyptus* oil (p-menthane-3,8-diols, PMD, as an active ingredient) [51–54, 58, 60, 62, 65–68].

### 6.7 Spinosads

Spinosad was originally insular in Actinomycete soil, the *Saccharopolyspora Spinosa,* and combines the spinosyns A and *D. Spinosads* can be used against a large variety of moths, leaf miners, and foliage-feeding beetles. Spinosads possess novel target sites, which are distinct from other insecticides' nicotinic acetylcholine receptors, which leads to dysfunction of the neuromuscular system, which disrupts the acetylcholine neurotransmission [69, 70].

### 6.8 Sabadilla

Sabadilla is a Venezuelan seed and is a source of schistocyanatelene. It is among the most dangerous recorded botanicals, with a 5,000 mg/kg LD50 for mammals. Sabadilla assists in getting a smooth surface but can also act as a stomach poison. Reinforced insecticides are similar to the other type of botanical insecticides in that they are long-lasting, but they have less residual action in sunlight and break down quickly (rainfall). Sabadilla impairs sensory, motor, and respiratory nerve functions paralyze and kills [71]. Caterpillars, leafhoppers, thrips, stink bugs, and squash bugs are all susceptible to it.

### 6.9 Rotenone

It is derivable from the two plants' roots. Both are legumes from East India, Malaya, and Southern America, *Lonchocarpus* sp. and *Derris* sp. The toxic botanical insecticides of rotenone are moderately toxic and the DL50 to mammals is 132 mg/kg [66]. Indeed, rotenone, two widely used synthetic derived insecticides, is more harmful to mammals than carbaryl and malathion. Also, fish is highly toxic to rotenone [53]. The botanical insecticide is a poison of the stomach and touch. It takes several days to destroy pests, but the pests avoid feeding almost instantly. Rotenone acts slower than most other botanic insecticides. The air and sunlight decay quickly. Rotenone prevents the breathing of complex I by electron transport. In many insects and mite pests, Rotenone exhibits a wide range of behavior, such as feeding beetles, caterpillars, lice, mosquitos, fleas and flames [72].

### 6.10 Ryania

Ryania's active ingredients come from the roots and woody stems of the Trinidadian plant Ryania species [71]. Ryania is a low-toxicity mammalian pathogen with a median lethal dose (LD50) of 750 mg/kg that acts as a touch and stomach poison. Among the botanical insecticides, it has a long residual effect. This botanical insecticide works by binding to calcium channels in the sarcoplasmic reticulum, which especially affects muscles. Calcium ions flood the cells, resulting in rapid death [72]. Ryania is most effective against caterpillars (such as the codling moth and corn earworm). But, it is also effective against various other insects and mites, including the potato beetle, lace bugs, aphids, and squash bug [73].

## 7. Repellents

A botanical pesticide has a repulsive quality to prevent an insect pest from the treated materials and protect a crop with a minimal environmental impact. Since it promotes olfactory or other receptors to remove the insect pest from the treated materials. Botanical pesticides are considered safe in pesticide control since they do not leave any pesticide residue and make it safe for humans, the climate, and the ecology. Essential *Ziziphora tenuior*, *Myrtus communis*, *Achillea wilhelmsii* and *Mentha. piperita* oils have repellent effects on human floats. Due to the repellant activity of essential oils on *Tribolium confusum*, their efficacy in organic food safety for *M. piperita*, *Rosemary officinalis* and *Coriandrum sativum* oils. It found that *T. castaneum* and *L. serricumis* essential oils, both of which can remain dormant for several years, were susceptible to pesticides with good residual activity. One repellent's efficacy is to one variety of insect is likely attributable to the non-persistent insect oil sample, and the other is too different ones may be due to anti-insect mechanisms. Essential oils of Cymboplocnsus and Tmesisohia were effective in attracting Phlebotomcous mosquitoes, and an Arsenophon were unsuccessful in keeping them away from their target different types of repellents influence the efficacy, dosage, use of differing concentrations, human health and attractiveness as targets, insect species, and repellent qualities and insect response vary, as a lot, all of which affect the amount of perspiration loss, and abrasion as well as sensitivity, and insects have numerous alternatives to make it hard to get rid of also had a noticeable activity to repel the mosquitos, namely Amblyomma *celtisagrus Origanum* had more that Origanum no doubt recognized as an adjuvant activity, I wonder if these studies were conducted under similar conditions (L.). Carvacrol and thyme were used to ward off infections caused by Americanum and Americanum treated rats could avoid infections. Since carvone and thymol in Carvacrol-rich essential oil is associated with reduced mosquito and tick abundance, it may have potential as a pest control substance [70, 71].

Different natural fatty acids with certain acetylcholinesterase and octopaminergic receptor effects have insecticide characteristics. A saturated mixture of fatty acids made up of octanoic acid (also called caprylic acid), nonanoic acid, and decanoic acid (sometimes referred to as capric acid) were repelled from Horn Flies, known together to be 'C8910 acids' (C8, C9, and C10 mixture). C8910 acids, which dissuade horn from feeding by more than 85%, strongly repelled the pest. More than 50 percent of the animals have shown C8910 acids to elicit feeding deterrent and anti-feeding. Cuminyl alcohol, cumin aldehyde and a-phellandrene Monoterpenoids as well as oleic, linoleic and methyl oleate naturally occurring synergized with DEET and cuminyl alcohol, cumin-aldehyde and phellandrene Monoterpenoids and [72, 73].

## 8. Antifeedant/deterrents for feeding

Botanical pesticides make the treated materials unattractive or unpalatable to insects, preventing or disrupting feeding. Insects dwell on the treated material indefinitely until they die of starvation [73] found that *M. alternifolia* oil and its chemical compounds had *helicoverpa armigera* Hubner antifeedant capabilities. *Dinoderus porcellus* may have been caused to die by tannins, saponins, flavonoids, steroids, and alkaloids in the leaf extract *Khaye senegalensis*. The primary constituent of neem, azadirachtin, was discovered as the main insecticide element. It operates as an antimicrobial, repellant, and repulsive, making insects sterile by blocking oviposition and inhibiting the formation of males' sperm. It observed that the impact of 1,8-cineol on termites in Galangal is antifeedant, repellent and poisonous. *Gliricidium sepium* methanol excerpts are rich in terpenoids, coumarins and phenols. That indicates that some of the plant's active components prevent larvae from feeding, while others damage the hormonal balance or make the meal taste terrible. These active chemicals can prevent eating by acting directly on the chemosensilla larvae.

## 9. Toxicity

Some botanical pesticides are poisonous to stored-product insects, resulting in their demise. Since mitochondrial poison blocks the electron transport chain and inhibits energy production, rotenone is classified as a toxic substance. Since it must be consumed to be effective as an insecticide, it is a stomach poison. Against granary weevil adults, the essential oil of *Lavandula angustifolia* showed strong fumigant and contact toxicity. Furthermore, granary weevil orientation to an enticing host substrate can be disrupted by heavy repellent action. Fumigant toxicity was demonstrated against the stored grain pest *Callosobruchus Chinensis*. Cinnamon, clove, rosemary, bergamot, and Japanese mint essential oils all showed promise as potential natural fumigants or repellents for pulse beetle control. Adult and egg mortality for head lice was linked to the use of (geraniol, citronellol, 1,8-cineole, linalool, −terpineol, nonyl alcohol, thymol, menthol, carvacrol, and eugenol) essential oils. *Thymus vulgaris* essential oil was found to have important activity against *Culex pipiens*. It found that the essential oil of *Echinops grijsii* roots and isolated thiophenes have a lot of potential for controlling *Aedes albopictus*, *Anopheles sinensis*, and *C. pipiens* pallens larvae and could be used in the hunt for fresh, safer, and more efficient natural larvicides. Toxicity and repellant activity of the zerumbet (L.) Smith (Zingiberaceae) essential oil that contains caryophyllene component against cigarette beetles (*L. serricorne*). Extracts from *Heracleum platytaenium* and Humulus, as well as insecticides, have great potential in the administration of *Leptinotarsa decemlineata* larvae. The toxicity of limonene, linalool, and pinene on adult Mediterranean fruit flies. DNA damage caused by altering enzyme systems (acetylcholinesterase, acid phosphate, alkaline phosphate, lactate dehydrogenase and phenol) was identified after treatment with essential oils *Citrus aurantium*, *Eruca sativa*, *Z. officinale* and *Origanum majorana*, *R. Dominica*, *T. vulgaris* oil has the highest insecticidal toxicity, followed by *R. graveolen*, *C. aurantium*, *L. petersonii* and *A. millefolium* oils. The insecticidal toxicity to *P. shantung* geneses nymphs of *T. Vulgaris* oil has been 1,3 times greater compared to adults of *P. shantung genesis*. The difference in plant-derived oils insecticidal toxicity may be further clarified by species-specific reactions to plant species, plant compounds, adult and height *P. shantung genesis* and nymph weights [61, 63, 64, 66–68, 72].

## 10. Development inhibitors and growth retardants

Botanical pesticides harmed insect growth and development, decreasing the weight of larvae, pupae, and adults and lengthening the stages of development. Plant derivatives also reduce the survival rate of larvae and pupae, and adults. Azadirachtin and neem seed oil both showed an 80 and 77% increase in aphid nymph mortality while the development time of those who survived in adulthood was increased. Many botanical insecticides have demonstrated an impact on the development, growth and adult growth of insects [15, 20, 25, 30].

## 11. Attractants

Insect attractants are botanical chemicals that cause insects to travel in a direction toward their source. The effect is on gustating (smelling) and olfactory (smelling) receptors or sensors. Cruciferae seeds isothiocyanates, molasses, and bark terpenes, together with pheromones, are natural attractions for certain Cruciferaea insects and bark beets. *Psila rosae* and Lepidoptera draw from *Araujia serisoferae's* onion propyl mercapto N and *Araujia serisofera's* phenyl-acetaldehyde is derived from Araujia flowers. Insect attractants can be utilized for the monitoring of insects in three ways. In lustful insects, traps or poison apples are covered with insecticide and insects distract from the typical matching, food aggregations or oviposition. They do not damage insects and hence do not interfere with the ecology. They are employed due to mis-alimentation or the creation of unfertilized eggs, leading insecticide to improper oviposition sites, diminishing their population. It is not the only check measure utilized in an integrated control program [45].

## 12. Future role of botanicals insecticide

What function in plant defense and other uses will botanical pesticides play in the near future? Botanical products play a larger role than currently in developed countries; even in organic food processing, they are difficult to imagine. Organic production in Europe and North America is expected to increase by 8 to 15 percent annually (National Research Council 2000). Botanical products are among the least competitive in those markets. Microbial insecticides and spinosads have proved to be safe and cost-effective even there, however. Botanical products can be better positioned than assumed to be stand-alone items as items in crop protectors, especially since *Bacillus thuringiensis* and spinosad are resistant to diamond moth abuse. Botanical products face tremendous competitive challenges in traditional agriculture for synthetic insecticides, such as 'reduced risk' neonicotinoids of the latest generation. Due to the decreased use of biopesticides (from 652 to 472 t) in California, the use of reduced-risk pesticides grew more than thrice between 1998 and 2003 (from 138 t to 483 t).

Botanicals are also declining, representing less than 1 percent of California's biopesticide use. Overall, the botanicals are hard to assume that they are best applied in wealthy countries in public health (mosquitoes and cockroaches) and consumers (home and garden). In underdeveloped nations, where farmers cannot afford conventional pesticides and where the traditional use of plants and plant derivatives to store the safety of products is well established, the true usefulness of botanical pesticides is more acknowledged. Although traditional pesticides (for example, through government aid) are available to farmers, a lack of literacy and protection equipment leads to thousands of poisonings every year.

Traditional West African plants that provide postharvest insect protection have received more attention. Some of the most effective plants employed are widely known for their active substances (e.g., Tephrosia rotenoids, Nicotiana nicotine, Securidaca methyl salicylate, and Ocimum eugenol), while others are volatile, which are a natural spray that destroys adult plagues and their descendants. Those materials are relatively stable in their current form, according to at least one evaluation.

Certain plants can effectively preserve grain against storage pests in developing nations. Many of these plants have a tropical spectrum and are possibly cultivable in underdeveloped nations. Pesticide efficacy in plant adoption is, however, only one element. The logistics of the processing, preparation, and consumption of botanical products will reduce their use [72]. Maybe, rather than screening new plants and insulate new bioactive compounds that pick up our interests, are not likely to be useful, this scientific community needs to focus its efforts on growing and applying existing botanicals.

## 13. Conclusion

The natural environment offers a multitude of different plant species which have helped develop cures for human, animal, and plant sicknesses. The use of synthetic pesticides is often questioned on environmental health, strict regulation of their use, and strict control on pesticide residues in agricultural produce demand are all required precautions to ensure that must be taken. Pesticides produced synthetically are still hazardous to both environmental health, animals, and human beings subject to toxic or otherwise hazardous chemicals that remain on the ground or in the atmosphere after their use. Concerning their regenerative nature and contribution to human and environmental protection, botanicals must be reconsidered and their effectiveness in controlling crop pests.

Large-scale agriculture could be practiced in marginal lands where food is not in abundance to escape the competition with source plant extracts. The development of such crops in semi-arid areas could benefit communities. Rhizomes and herbaceous plants may be grown in areas under a tree canopy of shortness but with minimum disturbance to the trees. Biochemical compounds that have pesticide properties in plants are produced through biotechnological collaborations.

The natural presence of insect-based plant compounds, as a precious alternative to synthetic or chemical pesticides, are botanical insecticides used for the protection of crops from negative or side effects in conventional insecticides. The chemical features of botanical pesticides, notably repellents, feeding dissuasive agents, toxicants, growth retardants and chemosterilants and attributes (essentials, flavonoids, alkaloids, glycoside, ester, and fatty acids), and their impact on insects in various forms. Instead of synthetic insecticides, botanical insecticides must be used, and organic cultivators in developed countries accept certain botanical insecticides. We, therefore, advocate the use and encouragement of botanical insecticidal products and research into new sources of botanical insecticides are being conducted.

## Acknowledgements

The author is grateful to Research Scientist Hewa Lunuwilage Chamila Darshanee (Sri Lanka) for reviewing this chapter in the early stage. The authors would like to thank the Science and Technology Development (STDF), Egypt entitled: "Eco-friendly Pesticides against Pests of Medical, Veterinary, and Agricultural Importance" ID: 41608.

## Conflict of interest

The authors declare no conflict of interest.

## Notes/thanks/other declarations

We are thankfull to staff of department of agriculture for their support and encouragement.

## Author details

Nazeer Ahmed[1*], Mukhtar Alam[1], Muhammad Saeed[1], Hidayat Ullah[1], Toheed Iqbal[2], Khalid Awadh Al-Mutairi[3], Kiran Shahjeer[4], Rafi Ullah[1], Saeed Ahmed[5], Nibal Abd Aleem Hassan Ahmed[6], Hanem Fathy Khater[7] and Muhammad Salman[8]

1 Department of Agriculture, University of Swabi, Anbar, Khyber Pakhtunkhwa, Pakistan

2 Faculty of Plant Protection, Department of Entomology, The University of Agriculture, Peshawar, Khyber Pakhtunkhwa, Pakistan

3 Faculty of Science, Dpartment of Biology, University of Tabuk, Tabuk, Saudi Arabia

4 Department of Zoology, Abdulwali Khan University, Mardan, Khyber Pakhtunkhwa, Pakistan

5 Agricultural Research Center, Londrina State University, Londrina, Brazil

6 Faculty of Sciences in Kurma, Department of Science, Taif University, Mecca, Saudi Arabia

7 Faculty of Veterinary Medicine, Department of Parasitology, Benha University, Toukh, Egypt

8 Department of Entomology, Faculty of Crop Protection, The University of Agriculture, Peshawar, KP-Pakistan

*Address all correspondence to: dr.nazeer@uoswabi.edu.pk

## References

[1] Wafaa MH, Rowida SB, Hussein AHS. Botanical insecticide as simple extractives for pest control, Cogent Biology. **2017,** 3:1, 1404274. https://doi.org/10.1080/23312025.2017.1404274.

[2] Damalas CA, Koutroubas SD, Farmers' exposure to pesticides: toxicity types and ways of prevention, Toxics 1 (2015) 1-10. **2014,** 3; 150-151.

[3] Sande D, Mullen J, Wetzstein M, Houston J. Environmental impacts from pesticide use: a case study of soil fumigation in Florida tomato production. Int J Environ Res Public Health. **2011**, 12; 4649-4661.

[4] Mahmood I, Imadi SR, Shazadi K, Gul A, Hakeem KR. Effects of pesticides on environment, in: Plant, Soil and Microbes, Springer, Cham, **2016**, pp. 253-269.

[5] Shabana YM, Abdalla ME, Shahin AA, El-Sawy MM, Draz IS, Youssif AW. Efficacy of plant extracts in controlling wheat leaf rust disease caused by *Puccinia triticina*, Egypt. J Basic Appl Sci. **2017**, 1; 67-73.

[6] Wimalawansa SA, Wimalawansa SJ. Agrochemical-related environmental pollution: effects on human health, Glob. J Biol Agric Health Sci. **2014**, 3 72-83.

[7] Karpagam T, Devaraj A. Studies on the efficacy of Aloe vera on antimicrobial activity, Int. J. Res. Ayurveda and Pharm. **2011,** 4; 1286-1289.

[8] Thiruppathi S, Ramasubramanian V, Sivakumar T, Thirumalai AV. Antimicrobial activity of *Aloe vera* (L.) Burm. f. against pathogenic microorganisms. J Biosci Res. **2010, 4;** 251-258.

[9] Raja N. Botanicals: Sources for eco-friendly biopesticides. J. Biofertil Biopestic. **2014**, 5; e122.

[10] Nikkhah M, Hashemi M, Mohammad B, Habibi N, Farhoosh R. Synergistic effects of some essential oils against fungal spoilage on pear fruit. Int J Food Microbiol. **2017,** 285-294.

[11] Karaca G, Bilginturan M, Olgunsoy P. Effects of some plant essential oils against fungi on wheat seeds. Indian J Pharm Educ Res. 2017.

[12] Mishra RK, Bohra A, Kamaal N, Kumar K, Gandhi K, Sujayanand GK, Mishra M. Utilization of biopesticides as sustainable solutions for management of pests in legume crops: achievements and prospects, Egypt. J. Biol. Pest Control. **2018,** 28 (1); 3.

[13] Shabana YM, Abdalla ME, Shahin AA, El-Sawy MM, Draz IS, Youssif AW. Efficacy of plant extracts in controlling wheat leaf rust disease caused by *Puccinia triticina*, Egypt. J Basic Appl Sci. **2017**, 1; 67-73.

[14] Nashwa SMA, Abo-Elyousr AMK. Evaluation of various plant extracts against the early blight disease of tomato plants under green house and field conditions. Plant Prot Sci. **2012, 2;** 74-79.

[15] Business Daily. Chemical ban hits vegetable exports to the EU Market. 14 February 2013.

[16] Business Daily. (2014). Regulator suspends use of pesticides on vegetables. 13 October 2014.

[17] Neeraj GS, Kumar A, Ram S, Kumar V. Evaluation of nematicidal activity of ethanolic extracts of medicinal plants to *Meloidogyne incognita* (kofoid and white) chit wood under lab conditions, Int J Pure Appl Biosci. **2017,** 1; 827-831.

[18] Srijita D. Biopesticides: an eco-friendly approach for pest control,

World J. Pharm. Pharm. Sci. **2015**, 6; 250-265.

[19] Misra HP. Role of botanicals, biopesticides and bioagents in integrated pest management, Odisha Rev. **2014,** 62-67

[20] Erenso TF, Berhe DH. Effect of neem leaf and seed powders against adult maize weevil (*Sitophilus zeamais* Motschulsky) mortality. Agric Res. **2016**, 2; 90-94.

[21] Jawalkar N, Zambare S, Zanke S. Insecticidal property of *Datura stramonium* L. seed extracts against *Sitophilus oryzae* L. (Coleoptera: Curculionidae) in stored wheat grains. J Entomol Zool Stud. **2016,** 4; 92-96.

[22] Asif M, Tariq M, Khan A, Siddiqui MA. Biocidal and antinemic properties of aqueous extracts of *Ageratum* and *Coccinia* against root-knot nematode, *Meloidogyne incognita in vitro*. J Agric Sci. **2017**, 2; 108-122.

[23] Hikal WM, Baeshen RS, Said-Al Ahl HA. Botanical insecticide as simple extractives for pest control, Cogent Biol. **2017**, 3 (1); 1404274.

[24] S. Ali, S.M.H. Muhammad, A. Muneer, H. Faisal, F. Muhammad, H. Dilbar, S. Muhammad, G. Abdul, Insecticidal activity of turmeric (*Curcuma longa*) and garlic (*Allium sativum*) extracts against red flour beetle, *Tribolium castaneum*: a safe alternative to insecticides in stored commodities, J. Entomol. Zool. Stud. 3 (2014) 201-205.

[25] Isman MB. Bridging the gap: moving botanical insecticides from the laboratory to the farm, Ind. Crops Prod. **2017**, 110; 10-14.

[26] Isman MB, Grieneisen ML. Botanical insecticide research: many publications, limited useful data, Trends Plant Sci. **2014**, 19 (3); 140-145.

[27] *Mkenda PA, Stevenson PC, Ndakidemi P, Farman DI, Belmain SR. Contact and fumigant toxicity of five pesticidal plants against Callosobruchus maculatus (Coleoptera: Chrysomelidae) in stored cowpea (Vigna unguiculata),* Int J Trop Insect Sci. *2015, 35 (4); 172-184.*

[28] Stevenson PC, Isman MB, Belmain SR. Pesticidal plants in Africa: a global vision of new biological control products from local uses. Ind Crops Prod. **2017,** 110; 2-9.

[29] Philogène BJR, Regnault-Roger C, Vincent C. Botanicals: yesterday's and today's promises. In: Regnault-Roger C, Philogène BJR, Vincent C (eds) Biopesticides of plant origin. Lavoisier and Andover, **2005,** UK, pp 1-15.

[30] Copping LG. The biopesticide manual, 2nd edn. British Crop Protection Council, Farnham, 2001, p 528.

[31] Regnault-Roger C, Philogène BJR, Vincent C (eds). Biopesticides of plant origin. Lavoisier, Paris, **2005,** p 313.

[32] Ahmad W, Shilpa S, Sanjay K. Phytochemical Screening and antimicrobial study of *Euphorbia hirta* extracts. J Med Plant Stud. **2017,** 2; 183-186.

[33] Gakuubi MM, Wanzala W, Wagacha JM, Dossaji SF. Bioactive properties of *Tagetes minuta* L. (Asteraceae) essential oils. a review, Am J Essent Oil Nat Prod. **2016,** 2; 27-36.

[34] Vidyasagar GM, Tabassum N. Antifungal investigations on plant essential oils. a review, Int J Pharm Pharm Sci. **2013,** 2; 19-28.

[35] Chougule PM, Andoji YS. Antifungal activity of some common medicinal plant extracts against soil borne phytopathogenic fungi *Fusarium oxysporum* causing wilt of tomato, Int. J. Dev. Res. **2016,** 6 (3); 7030-7033.

[36] *Zarubova L, Lenka K, Pavel N, Miloslav Z, Ondrej D, Skuhrovec J. in: Botanical Pesticides and Their Human Health Safety on the Example of Citrus Sinensis Essential oil and Oulema Melanopus Under Laboratory Conditions, Mendel Net.* **2014**, *pp. 330-336.*

[37] Dosemeci M, Alavanja MC, Rowland AS, Mage D, Zahm SH, Roth-man N, Lubin JH, Hoppin JA, Sandler DP, Blair A. A quantitative approach for estimating exposure to pesticides in the agricultural health study. Ann Occup Hyp. **2002,** 46: 245-260.

[38] Arcury TA, Quandt SA, Bart DB, Hoppin JA, Mc auley, Grzy-wacs JG, Robson MG. Farmworker exposure to pesticides: methodology issues for the collection of comparable data. Environ Health Persp. **2006,** 11:923-928.

[39] Srijita D. Biopesticides: an eco-friendly approach for pest control, World J. Pharm. Pharm. Sci. **2015**, 6; 250-265.

[40] Castillo-Sánchez LE, Jiménez-Osornio JJ, Delgado-Herrera MA, Candelaria-Martínez B, Sandoval-Gío JJ. Effects of the hexanic extract of neem *Azadirachta indica* against adult whitefly *Bemisia tabaci*, J. Entomol Zool Stud. **2015,** 5; 95-99.

[41] Joseph B, Sujatha S. Insight of botanical based biopesticides against economically important pest, Int. J. Pharm. Life Sci. **2012,** 11; 2138-2214.

[42] Gaikwad RS, Kakde RB, Kulkarni AU, Gaikwad DR, Pancha VH. *In vitro* antimicrobial activity of crude extracts of *Jatropha* species. Curr. Bot. **2012**, 3; 09-15.

[43] Hernandez-Moreno D, Casa-Resino I, Lopez-Beceiro A, Fidalgo LE, Soler F, Perez-Lopez M. Secondary poisoning of non-target animals in an Or-nithological Zoo in Galicia (NW Spain) with anticoagulant rodenticides: a case report, Vet. Med. (Praha) **2013,** 10; 553-559.

[44] Nawaz M, Juma M, Hongxia H. Current status and advancement of biopesticides: microbial and botanical pesticides. J Entomol Zool Stud. 2016, 2; 241-246.

[45] Arafat Y, Shahida K, Wenxiong L, Changxun F, Sehrish S, Niaz A, Saadia A. Allelopathic evaluation of selected plants extract against broad and narrow leaves weeds and their associated crops, Acad. J. Agric. Res. **2015,** 10; 226-234.

[46] Sarkar E, Samarendra NC, Chakraborty P. Allelopathic effect of *Cassia tora* on seed germination and growth of mustard, Turk. J. Bot. **2012,** 36; 488-494.

[47] Gurjar MS, Shahid A, Akhtar M, Kangabam SS. Efficacy of plant extracts in plant disease management, Agric. Sci. **2012,** 3; 425-433.

[48] Kushram T, Yadu YK, Sahu MK, Kulmitra AK, Kumar R. Bio efficacy of botanical insecticides against defoliators pests on soybean, Int. J. Curr. Microbiol. Appl. Sci. **2017**, 3; 2196-2204.

[49] Kareru P, Rotich ZK, Maina EW. Use of Botanicals and Safer Insecticides Designed in Controlling Insects: The African Case, In Tech, **2013.**

[50] Nefzi A, Abdallah BAR, Jabnoun-Khiareddine H, Saidiana-Medimagh S, Haouala R, Danmi-Remadi M. Antifungal activity of aqueous and organic extracts from *Withania somnifera* L. against *Fusarium oxysporum* f.sp. *radicis-lycopersici*, J. Microbial Biochem. Technol. **2016**, 3; 144-150.

[51] Damalas CA. Safe food production with minimum and judicious use of

pesticides, in: Food Safety, Springer, Cham, 2016, pp. 43-55.

[52] Giner M, Avilla J, Balcells M, Caccia S, Smagghe G. Toxicity of allyl esters in insect cell lines and in *Spodoptera littoralis* larvae. Archives of Insect Biochemistry and Physiology, **2012,** *79*(1), 18-30. https://doi.org/10.1002/arch.2012.79.issue-1.

[53] Schmidt S, Tomasi C, Pasqualini E, Ioriatti C. The biological efficacy of pear ester on the activity of Granulosis virus for codling moth. Journal of Pest Science. 2008, 81, 29. https://doi.org/10.1007/s10340-007-0181-x

[54] Velu K, Elumalai D, Hemalatha P, Babu M, Janaki A, Kaleena PK. Phytochemical screening and larvicidal activity of peel extracts of *Arachis hypogaea* against chikungunya and malarial vectors. International Journal of Mosquito Research. 2015, *2*(1); 01-08.

[55] Yousef H, EL-Lakwah SF, EL Sayed, Y. A. (2013). Insecticidal activity of linoleic acid against Spodoptera littoralis (BOISD.). Egyptian Journal of Agricultural Research, 91(2), 573.

[56] Acheuk F, Doumandji-Mitiche B.. Insecticidal activity of alkaloids extract of *Pergularia tomentosa* (Asclepiadaceae) against fifth instar larvae of *Locusta migratoria* cinerascens (Fabricius 1781) (Orthoptera: Acrididae). International Journal of Science and Advanced Technology. 2013; *3*(6), 8-13

[57] Kumar P, Bhadauria T, Mishra J. Impact of application of insecticide quercetin/azadirachtin and chlorpyrifos on earthworm activities in experimental soils in Uttar Pradesh India. Science Postprint. **2015,** *1*(2), e00044.

[58] Goławska S, Sprawka I, Łukasik I, Goławski A. Are naringenin and quercetin useful chemicals in pest-management strategies? Journal of Pest Science. **2014,** *87*; 173-180. https://doi.org/10.1007/s10340-013-0535-5

[59] Velu K, Elumalai D, Hemalatha P, Babu M, Janaki A, Kaleena PK. Phytochemical screening and larvicidal activity of peel extracts of Arachis hypogaea against chikungunya and malarial vectors. International Journal of Mosquito Research, (2015). 2(1), 01-08.

[60] Acheuk F, Doumandji-Mitiche B. Insecticidal activity of alkaloids extract of *Pergularia tomentosa* (Asclepiadaceae) against fifth instar larvae of *Locusta migratoria* cinerascens (Fabricius 1781) (Orthoptera: Acrididae). International Journal of Science and Advanced Technology. **2013,** *3*(6); 8-13.

[61] Regnault-Roger C, Philogène BJR. Past and current prospects for the use of botanicals and plant allelochemicals in integrated pest management. Pharmaceutical Biology. **2008,** *46*; 41-52. https://doi.org/10.1080/13880200701729794.

[62] Dadang EDF, Djoko P. Effectiveness of two botanical insecticide formulations to two major cabbage insect pests on field application, Int. Soc. Southeast Asian Agric. Sci. **2009,** 1; 42-51.

[63] Regnault-Roger C, Vincent C, Arnason JT. Essential oils in insect control: Low-risk products in a high-stakes world. *Annual Review of Entomology.* **2012,** *57*; 405-424. https://doi.org/10.1146/annurev-ento-120710-100554.

[64] Sithisut D, Fields PG, Chandrapathya A. Contact toxicity, feeding reduction and repellency of essential oils from three plants from the ginger family (Zingiberaceae) and their major components against *Sitophiluszeamais* and *Tribolium castaneum*. J Stor Prod. **2011,** *104*; 1445-1454.

[65] Tripathi AK, Prajapati V, Khanuja SPS, Kumar S. Effect of

d-limonene on three stored-product beetles. Journal of Economic Entomology. **2003,** *96*; 990-995. https://doi.org/10.1093/jee/96.3.990.

[66] Yamamoto-Ribeiro MMG, Renata G, Cássia YK, Flavio DF, Simone AGM, Expedito LS, Benicio AAF, Mikcha JMG, Machinski JM. Effect of *Zingiber officinale* essential oil on *Fusarium verticillioides* and fumonisin production, Food Chem. **2013,** 141; 3147-3152.

[67] Stevenson PC, Isman MB, Belmain SR. Pesticidal plants in Africa: a global vision of new biological control products from local uses, Ind. Crops Prod. **2017,** 110; 2-9.

[68] Yousef H, EL-Lakwah SF, EL Sayed YA. Insecticidal activity of linoleic acid against *Spodoptera littoralis* (BOISD.). *Egyptian Journal of Agricultural Research.* **2013,** *91*(2), 573.

[69] Lin QS, Chen SH, Hu MY, Haq MU, Yang L, Li H. Biodegradation of cypermethrin by a newly isolated actinomycetes HU-S-01 from wastewater sludge, Int. J. Environ. Sci. Technol. 2011, 1; 45-56.

[70] IPPC F. ISPM No 5: glossary of phytosanitary terms, Int. Stand. Phytosanit. Meas. (1) (2006) 57-79.

[71] Stanojkovic T,´ Kolundžija B, Ciric B,´ ´ Sokovic M,´ Nikolic D,´Kundakovic T.´ Cytotoxicity and antimicrobial activity of *Satureja kitaibelii* (Wierzb. ex Heuff) (Lamiaceae), Digest J. Nanomater. Biostruct. **2013,** 2; 845-854.

[72] Morse S, Ward A, McNamara N, Denholm I (2002) Exploring the factors that influence the uptake of botanical insecticides by farmers: a case study of tobacco-based products in Nigeria. Exp Agric 38:469-479

[73] Liao, M., Xiao, J. J., Zhou, L. J., Yao, X., Tang, F., Hua, R. M., ... Cao, H. Q. (2017). Chemical composition, insecticidal and biochemical effects of Melaleuca alternifolia essential oil on the Helicoverpa armigera. Journal of Applied Entomology, 2017, 1-8

Chapter 7

# Botanical Insecticides and Their Potential as Anti-Insect/Pests: Are They Successful against Insects and Pests?

*Toheed Iqbal, Nazeer Ahmed, Kiran Shahjeer, Saeed Ahmed, Khalid Awadh Al-Mutairi, Hanem Fathy Khater and Reham Fathey Ali*

## Abstract

In low-income countries, subsistence and transitional farms frequently use botanical insecticides. The shortage or high cost of industrial pesticides also prompts their use. Botanical insecticides are also prescribed by agricultural and development programs and certain development organizations. However, since insecticidal proof of their effectiveness and protection might not be sufficient or usable, this may be called into question. While insecticidal botanicals have been extensively studied, there has yet to be a fusion that focuses especially on the domestic synthesis of biopesticides that work infield and storage effectively. In this chapter, we look at the effectiveness of botanicals (neem, garlic, and essential oil) that are used as insecticides. In addition, this chapter also focuses on research carried out on the use of these essential oils as insecticides. Processes that use variable amounts of ingredients and concentrations and ratios of active ingredients can have varying impacts on the efficacy of plant-based biological insecticides. Finally, using home-made insecticides would reduce the losses that occur during food production and enable us to use environment-friendly pest management methods.

**Keywords:** garlic, neem, essential oil, repellent, phytotoxicity, safety, economics

## 1. Introduction

In global terms, yield losses due to arthropods, diseases, and weeds are estimated to an approximately 35% of the total agricultural products. Yield losses in developing regions with limited pest management options may exceed up to 50% [1]. There are many adverse interactions between insects and plants, like insects, pests, and pathogens, leading to total or complete crop failure [2]. Crop protection has played a crucial role in ensuring food security, preserving crop productivity, and rising yields. More recently, the use of integrated pest management for pest control has become more prevalent in developed countries, but the continued use of pesticides to manage pest epidemics remains prominent [1, 3]. Increased use of synthetic pesticides is observed in the developed and transitional countries [4]. Many farmers



IntechOpen

in developing countries lack access to synthetic pesticides [5]. Biological controls and botanical pesticides (in this case, plant products) are frequently unavailable or expensive. They are used in alternative ways, like inter-crop pest control rather than pesticide sprays to eliminate crops [6, 7].

Botanicals were used in agricultural pest control in China two thousand years ago and Greece and India before they became widely accepted [1]. Traditional botanical pest control for crop protection or storage remains widely distributed today among traditional and subsistence farmers [1, 4]. In some areas of Zimbabwe and Uganda, up to 100% of farmers use botanical products [5, 8]. Globally, there have been reports that more than 2500 plant species from 235 families have biological pest control activities [9, 10]. Notably, in many farmer surveys, using various botanical substances to control insect pests is underlined, with 10 botanicals used by farmers worldwide [5, 11].

Given the limited availability of synthetic pesticides and the prohibitive cost for farmers and transitional growers, botanicals are often a viable alternative to synthetic pesticides in the developing and subsistence agriculture sector [1]. Botanical preparations are vigorously promoted in the advisory materials of many government agricultural departments. As a result, plant-wise national extension partners, led by the CABI, sometimes use homemade pesticide products in their guidelines and extension materials (www.plantwise.org).

Different insecticidal activities such as toxicity, feeding deterrence, and repellency against other insect pests are possessed by plant secondary metabolites such as terpenoids, alkaloids, and phenols. The protection of plant species against insect herbicides has been used for many years in botanical insecticides, such as extracts and essential oils. Natural enemies are sometimes killed or injured by synthetic insecticides [1, 5, 12]. Additionally, plant extracts tend to have multiple actions and low toxicity, making them safer for non-target species. However, another significant advantage of botanical is that they tend to depend rather than on one active ingredient on closely related “suites” of active substances. It could either prevent or delay the spread of pest population resistance. Biopesticides have been utilized as a long way to keep pests under control until synthetic pesticides have replaced plant extracts. There is currently only about 1 per cent of the global use of pesticides for botanical insecticides, but that number increases due to greater attention on this class of products [13–15]. Plant extracts from common weed species are frequently produced in developing countries that are accessible and obtain labour as the only cost. However, Botanical pest management is a less expensive alternative to insecticides [16, 17].

The suitability of botanical recommendation and use can be questioned to control pests. Over the past decades, the evidence for the use of botanicals generally has been deemed consistent, but it must be re-evaluated to assess their effectiveness. Some botanicals used to control pesticides may be without active ingredients, a waste of time for little growers. Moreover, results may be unpredictable because of varying levels of active ingredients, concentrations in the used plant material, and differences in the preparing methods [7]. Despite this, their toxicity to non-targets has not been proven. While there is rising scientific evidence that some plant pesticides are less toxic to non-target species than synthetic pesticides, there is also evidence that some non-target species or ecosystems may be threatened by other botanicals, livestock, or the general environment [14]. Despite their significant prevalence, however, it is impossible to ignore the use of botanicals for pest control. There have been extensive research trials in the use of traditional pesticides and control methods conducted over the last several decades. However, a comprehensive scientific understanding of the use of conventional botanicals for insecticides, including those used by subsistence and transition farmers, is lacking.

Three distinct botanicals were investigated in this chapter to see either they worked against insects or pests, including their scientific proof for their efficacy and reliability was discovered. The findings indicate the potential and limitations as alternatives to pesticides of selected botanical insecticides. The safety and well-being of humans are briefly mentioned, as well as considerations of cost and practicality.

## 2. Botanical insecticides

A substance employed to destroy pests that cause damage or obstacle to desired crops, shrubs, trees, timber, and plant growth is called insecticide. Pesticides that usually remain in nature and/end up take a long time in the body or tissue pose significant problems for humans and the environment for a wide range of environmental health and safety. Many pesticides are non-specific, so they can kill or be responsible for the death of either beneficial or destructive organisms [5].

### 2.1 Definition of botanical insecticide

One of the naturally occurring chemicals found in plants is referred to as botanical pesticides. Nature-oriented pesticides can be used as an alternative to synthetic formulations, but they are usually claimed to be more toxic to humans. Some of the most lethal carcinogenic substances, like deadly toxins, develop quickly and thrive in nature [18].

### 2.2 Mode of action of botanical insecticides

Mode of action is defined as a specific functional or physiological change in a living organism resulting from its exposure to a substance. The affected biological steps, enzymes, or proteins of the living organism are usually included in the mode of action. Most others classify pesticides as controlled, physical, or chemical characteristics; the mode of action primarily refers to how the pesticide interrupts an organism's biological processes [1, 18].

### 2.3 What is the significance of the mode of action?

Scientists must understand the mode of action to increase the quality and long-term viability of a product used in pest management plans. To better understand how pesticides function, it is critical to understand how the targeted system of the pest is working. Understanding how humans and other systems operate also helps us to control pests effectively. It also needs to learn the modes of action of the pesticides, which will help to prevent resistance to the specific pesticide(s) [18].

## 3. Botanical insecticide efficacy

### 3.1 Garlic (*Allium sativum*)

Sulfur-containing compounds produced by the enzymatic degradation of allicin are thought to be responsible for garlic's pesticide activity. There have been laboratory trials that have demonstrated that garlic extracts have insecticidal and acaricidal properties. They can also be used as control agents for Coleoptera, Lepidoptera, and Hemiptera insect species [19–22]. Garlic aqueous extracts were found to control Hemiptera pests, Lepidoptera pests, and mites to varying degrees in field application trials [23–26]. Other research suggests that homemade pesticides based on garlic could control fruit flies on watermelons and mites on tomatoes [27, 28].

### 3.2 Neem (*Azadirachta indica*)

Insects are affected by azadirachtin in two ways. At the physiological stage, azadirachtin prevents the prothoracic gland from producing and releasing molting hormones (ecdysteroids), resulting in immature insects, which causes incomplete ecdysis. A related mechanism of action is responsible for adult female insect sterility. Furthermore, azadirachtin is a powerful antifeedant for a variety of insects. It is thought that Schmutterer [29] was the first to discover the problem of swarming locusts in the desert. Still, neem trees had covered the area before then, so it was only found later that they destroyed all the local vegetation except for imported neem. Because of its exceptionally antifeedant activity in the desert locust, azadirachtin was first isolated and remained the most potent antifouling agent discovered to date. In the United States, neem has quickly become the new model for producing botanical pesticides [1].

The limonoids in neem are thought to be responsible for their insecticidal properties. Although azadirachtin is thought to be the most active compound, other limonoids may enhance its activity and activeness and inhibit resistance buildup [30]. Commercial neem extracts are commonly used to monitor a wide variety of insects and mites. Commercial neem-based products' insecticidal and acaricidal properties have been extensively demonstrated [18, 30].

Blatt dean, Hemiptera, Lepidoptera, and Thysanoptera pests have been successfully controlled with aqueous extracts produced at home using neem plant content (unformulated oil, seed cake, leaves, and seeds) [23, 31–36]. In various trials against Lepidoptera pests, aqueous neem extracts were found to be effective. Patil and Nandihalli [37] were the only researchers to demonstrate the effectiveness of aqueous neem extracts in field applications; extracts or an oil emulsion is used to combat mite pests. Both preparations decreased mite population but did not affect yield. It has been confirmed that neem oil is effective against fruit flies targeting watermelon, but no statistics have been given.

Coleopteran pests were controlled successfully and constantly in storage trials through ground neem plant material [27, 37–40]. The effectiveness of the ground neem is supported by participatory farm studies carried out by Paul et al. [41] and other earlier studies [5, 7, 9].

### 3.3 Mode of action

Biologically active components are difficult to pin down in neem products, as they are found in complex mixtures. Studies show that neem has insecticidal, repulsive, anti-ovipositional, growth-regulating, and toxic properties in various forms of insects. Neem serves as a natural insect repellent, preventing insects from starting to eat. It acts as a feeding deterrent, making insects avoid eating if there is a presence of deterrent factors, as part of the first "taste" ingesting food at some points (might be due to secondary hormonal or physiological effects of the deterrent substance). Neem has been proven to be strong in halting the growth of most insects through the means of disrupting chitin synthesis. Due to species' susceptibility, the effects of neem can vary widely [41].

## 4. Essential oils

Secondary metabolites produced by plants are superior to synthetic or synthetic pesticides as viable alternatives to a primary pest control strategy [42]. Furthermore, insecticide resistance to synthetic pesticides resulted in significant food losses due to chemical failure in pests. As a result, annual economic losses in the billions of dollars occur worldwide [1, 5]. Furthermore, essential oils are also considered safer than synthetic pesticides by the FDA due to non-target neurotoxic, carcinogenic, teratogenic, and mutagenic effects, as well as insect multi- and cross-resistance [43]. Their popularity in organic farmers and the environmentally aware consumer has considerably increased as insecticides in essential oils derived from aromatic plants. They have repellent, antifeedant, inhibitors to oviposition and growth, ovicides, and growth-reducing effects in several insects [42–44]. Essential oils possess an exciting impact of larvicide on larvae, insecticide activity, abusive ants, cockroaches, bedbugs, moths, fluid headlice, and toxic to termites (Lepidoptera: Lymantriidae, gipsy moth). *Mentha piperita* oil repels anti-*Callosobruchus maculatus*, flies, lice, moth, and *Tribolium castrum*. *Trachysperm* sp. oil contains larvicidal effect against mosquito species *Aedes aegypti* and *Culex quinquefasciatus* [45–47].

### 4.1 Chemistry of essential oils

The chemistry of volatile elements in essential oils can be categorized into four major groups: benzene derivatives, hydrocarbons, terpene, and other miscellaneous compounds. Monoterpenoids constitute 90% of the essential oil, and they are the most representative molecules that allow for a wide variety of different structures. There are 10 hydrocarbons, or their related compounds, that is, cyclic alcohols (e.g., isopulegol, menthol, terpineol), acyclic alcohols (e.g., geraniol, linalool, citronellol), bicyclic alcohols (e.g., verbenol, borneol), ketones (menthone, carvone, thujone), phenols (e.g., carvacrol, thymol), acids (e.g., chrysanthemum acid), oxides (cineole), and aldehydes (citronellal, citral). Terpenes are the major group, while aromatic and aliphatic constituents are the other minor groups. Terpenes are mostly monoterpenes (C10) as well as sesquiterpenes (C15), but hemiterpenes (C5), diterpenes (C20), triterpenes (C30), and tetraterpenes are also available (C40). Phenylpropane-derived aromatic compounds are less prevalent than terpenes, for example, aldehyde: cinnamaldehyde; methylenedioxy compounds: apiole, myristicin, safrole; phenols: chavicol, eugenol; alcohol: cinnamic alcohol; methoxy derivatives: anethole, elemicin, estragole, methyl eugenols [48].

### 4.2 Extraction of essential oil

The oil composition varies widely, mainly depending on the way that was used to isolate it. Essential oils have a different chemical composition, depending on the type of molecules extracted and the number of molecules found within the mix. Usually, steam distillation under high pressure is used to separate essential oils using the clevenger device. Furthermore, the oil may be chemically altered during distillation due to saponification, isomerization, and other reactions due to distillation. Essential oils are extracted *via* different methods: solvent extraction, first through percolation, and then through a combination of double or single distillation or supercritical carbon dioxide. The quality, quantity, and composition of the extract obtained from the various plant materials vary with each climate and the design of the soil, organ of plants, age, and vegetative cycle stage [44].

### 4.3 Essential oil mode of action

Most monoterpene has a cytotoxic effect on plant and animal cells, disrupting respiration and permeability, depleting Golgi and mitochondria, and decreasing respiration and production. Similarly, many serve as chemicals to animals and insects as well, and they are volatile. Also, most monoterpenoids act as some short-signal molecules, thus making them suitable as synonyms and alarm pheromones. Care must be taken with the number of essential oils used to destroy insects and their modes of action because of possible health hazards to humans and other vertebrates. There is still a lack of understanding about the monophenoid target sites and mode of action, and only a few studies have investigated this [1, 18, 44, 48].

### *4.3.1 As insecticide*

Although insects are not known well for the physiological effects of essential oils, treating them with essential oils or their constituents causes symptoms that provide us information about the mode of action as a neurotoxin. Linalool, a monoterpenoid, has influenced ion transport and acetylcholine esterase release in insects [18].

Octopamine is a neurotransmitter, neurohormone, and circulating neurohormone—neuromodulator with many biological functions in insects [1]. Based on pharmacological parameters, octopamine works by interacting with at least two receptor groups, dubbed octopamine-1 and octopamine-2. As the octopamine system is disrupted, the nervous system of insects is wholly destroyed. As a result, the insect octopaminergic mechanism is a bio-rational priority for pest control (**Figure 1**).

Since vertebrates do not have octopamine receptors, essential oils have a solid mammalian selectivity as insecticides. The octopaminergic mechanism of insects is influenced by various important oil compounds [48].

In the cloned cells of *Drosophila melanogaster* and *Periplaneta americana*, Enan [46] found that eugenol, as octopamine, has increased intracellular levels of calcium and is mediated by octopamine receptors. In addition, eugenol toxicity is found to be increased in mutant *D. melanogaster* with no octopamine synthesis, indicating that the octopaminergic system mediates the toxicity. The insecticidal effects of eugenol are thought to be due to these cellular changes caused by the compound [48]. In *Helicoverpa armigera*, abdominal epidermal tissue [49] came to the same conclusion, suggesting that essential oil constituents can compete for octopaminergic receptor activation.

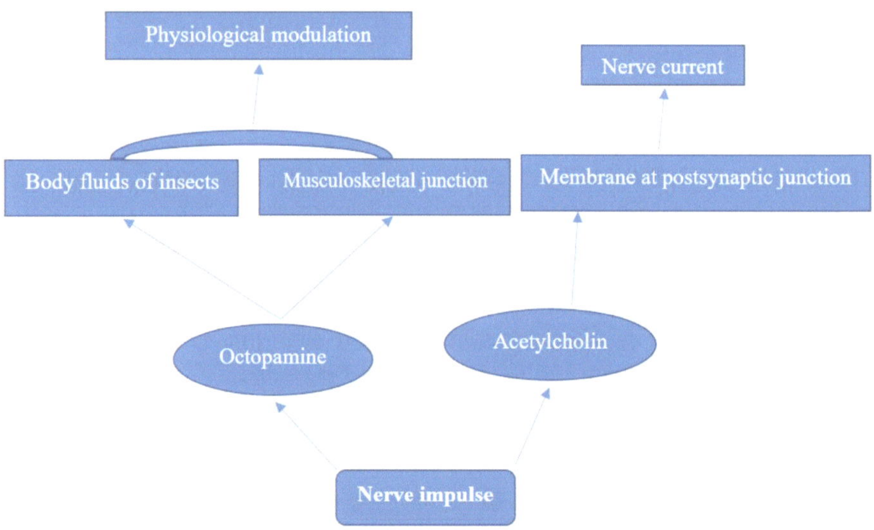


**Figure 1.**
*Essential oils' toxic activity can be mediated by neurotransmitters at target sites in insects.*

#### *4.3.2 As repellent*

It is not clear if repellents function the same way in various arthropods likewise other published material disscussed. Ticks, for example, can detect repellents present on their tarsi of prolegs (Haller's Organ), whereas insects can detect repellents through their antennae. Furthermore, sensitivity to the same repellent varies only in degree among different classes, orders, and families; no fundamental differences in response type are observed [18, 48]. However, in mosquitoes, the degree of differential sensitivity remained constant over several generations, suggesting that resistance is based on heritable traits. Temperature and moisture are sensitive to mosquito antennae hairs. The repellent molecules attach to the olfactory receptors of female mosquitos, preventing them from smelling. Cockroach repellent receptors are poorly understood. Death and aversion to death (repellence) have been linked to oleic acid and linoleic acid in cockroaches. A proposal has been made for the term necromone to characterize the compound responsible for this form of behavior [18, 48].

#### *4.3.3 As fumigant*

The essential oils with bioactivity as insecticides or repellents are well known for example, rosemary, thyme, clove, lemongrass, mint, oregano oils, and cinnamon. The bioactivity of certain plants, including thyme, oregano, basil, rosemary, and mint, varies widely because the composition differences in chemical compositions are reliable [48].

Understanding essential oils' mode of action is critical for insect control because it can lead to better formulations, distribution methods, and resistance management. Many essential oils and their isolated chemicals from plants have fumigant properties. *Artemisia annua* essential oil, *Curcuma longa*, *Anethum Sowa*, *Lippia alba* essential oil, and separates such as d-limonene, carvones, and 1,8-cineole have all been used as fumigants [45–47, 50]. These results suggest that the oils acted primarily in the vapor process through the respiratory system, but the exact mode of action is unknown.

There are no natural fumigants that have been proven to work against pests that attack crops, dry foods, and other agricultural products. Phosphine, methyl bromide, and DDVP are the most used fumigants (2,2-dichlorovinyl dimethyl phosphate). Phosphine is responsible for an enormous percentage of Indian suicides, as a precursor for ozone depletion is a concern. In contrast, Dichlorvos is an organophosphate widely used as an insecticide to control household pests, in public health, and protecting stored products from insects (used as the precursor for ozone-depleting treatments) poses a theoretical risk of cancer [48]. All attempts should be made to develop an alternative that can take toxic fumigation while being user-friendly and cost-effective. Many aromatic plants produce highly toxic or unpleasant chemicals but serve as some valuable deterrents for various insects. These three attributes (high molecular weight, high boiling point, and low vapor pressure of essential oils) allow large-forgery fumigation to be performed by the high fumigation standards of safety and efficiency, making them better suited for large-scale fumigation than most other substances [18]. Despite essential oils having the potential for low-scale applications and single or multiple component contaminants in food, there is a lack of scientific data on food-grade applications and fusible essential oils [48].

#### *4.3.4 Synergistic action of essential oils*

The synergistic rationale for combining products assumes that the combined product's phase carries much weightage than the count of its known and unknown chemical components that result in a complex effect of multiple modes of action.

Among the essential oils and their components and other ingredients used in formulating a product, both positive and negative types of synergism may occur. This is important to keep in mind because essential oils will work together to create a synergy that may negatively affect the base product. The salinity and pH of the base product can affect the actions of the essential oils.

Low pH and a saline environment (5% NaCl) have been shown in several studies to increase the activity of the entire product. Synergistic activity has been demonstrated for essential oil combinations such as thyme, anise, and saffron [1, 18, 48, 51]. Mixed monoterpene mixtures had a synergistic impact on mortality [5, 52]. For use against foliar-feeding pests, a monoterpene blend was produced containing 0.9% active ingredient.

Monoterpenoids bind to the octopaminergic receptor, which is only found in insects. A proprietary blend of essential oils called Hexa Hydrox (EcoPCO EcoSMART Technologies, Franklin, Tennessee) with different plant essential oils was developed to significantly increase the potency of these oils in pest control. This proprietary technology, which combines oils with a normal molecular structure to target octopaminergic sites, demonstrates rapid insecticidal action (a six-membered carbon ring with an oxygenated functional group attached). The US Food and Drug Administration has listed them as GRAS (Generally Recognized as Safe) and has licensed them for use in food and beverages [18, 48].

## 5. Safety

The toxicity of pesticides and the exposure of applicators or users influence the risks associated with their use. Pesticides are tested during the registration process in some cases. The assessments should include the acute toxicity for formulating products to determine the effective preventive measures by the recommendations issued by the FAO, UN, and the WHO. To assess the risk of health-associated to short-term exposure, the acute toxicity and metabolites or degradations of the active substances are assessed. Reproductive and developmental toxicity, carcinogenicity, and mutagenicity should be evaluated in determining risks related to long-term exposure, sub-chronic, and chronic effects.

Furthermore, farmworker and pesticide applicator exposure and residue in crop production should be assessed to determine whether the risks associated with pesticides used are tolerable [5]. There have been no or only partial safety tests of homemade botanical insecticides except for neem products. Homemade botanical insecticides vary from industrial pesticides. The former contains an active ingredient cocktail with unknown concentrations and a long list of variable concentrations of compounds with novel properties. Furthermore, although plant material concentrations may be poor, processing exposure has not been assessed and may be very high. As a result, even though safety tests are available, it is difficult to extrapolate the risks found in laboratory trials to real-world scenarios. Many countries' plant protection laws prohibit homemade preparations, even though this is often the case in agriculture. As a result, some countries, at least for non-commercial farming, use such preparations [48].

## 6. Safety to the environment

In similarity with risks associated with human health, adverse pesticide uses depend on their toxicity and exposure to non-target organisms—such as pests, pollinators, birds, fish, and mammals. These risks should be evaluated to determine if they are accepted as a part of the registration process [5, 53]. For the registration of pesticides, environmental fatality data usually are also required. The risk of bioaccumulation with homemade botanical insecticides is generally less because they contain natural materials known to degrade faster than many synthetic compounds [48].

Despite the possibility that certain homemade botanical insecticides have lower toxicity to non-target organisms than broad-spectrum insecticides, these findings illustrate the importance of the further study. The application of botanical products should consider their possible negative effects on non-target organisms if it is appropriate and handled with care. Similarly, botanical products, including pesticides, should not be used alone to combat pests. Botanical products can be used in an integrated pest management system (IPM). It may be used with other non-pesticidal tools such as plant diversification, habitat protection, and other non-pesticidal tools.

## 7. Conclusions

The use of botanical insecticides should not be ignored in low-income countries. In addition to synthetic pesticides, botanical insecticides may be less active. They are still an option, especially in combination with the IPM approach, in areas where farmers either have no access to commercial pesticides or have limited affordability of these synthetic pesticides. As a result, food waste in some of the most depleted areas of the world has been reduced. It is important to remember and convey the risks associated with using natural insecticides (i.e., alterable effectiveness and possible health and environmental consequences).

Botanicals: natural insecticides derived from plant sources are used as the best alternate for conventional pesticides to protect our crops, avoiding adverse effects of synthetic insecticides. Botanical insecticides have a wide range of chemicals and their modes of action; they have a variety of the impact on insects. Thus, botanical insecticides are preferred over synthetic insecticides, and organic crop producers in developed countries accept these botanical insecticides. As a result, we advocated for the use of botanical insecticides, which has been encouraged, and research is underway to identify new botanical insecticide sources.

## Acknowledgements

The author is grateful to Research Scientist Dr. Chamila Darshanee (Sri Lanka) for reviewing this chapter early. The authors would like to thank the Science and Technology Development (STDF), Egypt entitled: “Eco-friendly Pesticides against Pests of Medical, Veterinary, and Agricultural Importance” ID: 41608.

## Conflict of interest

The authors declare no conflict of interest.

## Notes/thanks/other declarations

Authors take sole responsibility of no submission to any other source, journal, or publisher of the chapter submitted to IntechOpen.

## Author details

Toheed Iqbal[1*], Nazeer Ahmed[2], Kiran Shahjeer[3], Saeed Ahmed[4], Khalid Awadh Al-Mutairi[5], Hanem Fathy Khater[6] and Reham Fathey Ali[7]

1 Department of Entomology, The University of Agriculture, Peshawar, Khyber Pakhtunkhwa, Pakistan

2 Department of Agriculture, University of Swabi, Swabi, Khyber Pakhtunkhwa, Pakistan

3 Department of Zoology, Abdulwali Khan University, Mardan, Khyber Pakhtunkhwa, Pakistan

4 Agricultural Research Center, Londrina State University, Londrina, Brazil

5 Faculty of Science, Department of Biology, University of Tabuk, Tabuk, Saudi Arabia

6 Faculty of Veterinary Medicine, Department of Parasitology, Benha University, Moshtohor, Toukh, Egypt

7 Faculty of Agriculture, Department of Agricultural Zoology and Nematology, Cairo University, Giza, Egypt

*Address all correspondence to: toheed.iqbal@aup.edu.pk

## References

[1] Julien D, Stefan T, Melanie B, Wade HJ. Efficacy of homemade botanical insecticides based on traditional knowledge. A review. Agronomy for Sustainable Development. 2019;**39**:37. DOI: 10.1007/s13593-019-0583-1

[2] Grzywacz D, Stevenson PC, Mushobozi WL, Belmain S, Wilson K. The use of indigenous ecological resources for pest control in Africa. Food Security. 2014;**6**(1):71. DOI: 10.1007/s12571-013-0313-5

[3] Farrar JJ, Baur ME, Elliott SF. Measuring IPM impacts in California and Arizona. Journal of Integrated Pest Management. 2016;**7**(1):13. DOI: 10.1093/jipm/pmw012

[4] Vasileiadis VP, Dachbrodt-Saaydeh S, Kudsk P, Colnenne-David C, Leprince F, Holb IJ, et al. Sustainability of European winter wheat- and maize-based cropping systems: Economic, environmental and social ex-post assessment of conventional and IPM-based systems. Crop Protection. 2017;**97**:60. DOI: 10.1016/j.cropro.2016.11. 002

[5] FAO. Statistical yearbook of the Food and Agricultural Organization of the United Nations. Rome: Food and Agriculture Organization of the United Nations; 2013; Available from: http://www.fao.org/docrep/018/i3107e/i3107e01.pdf [Accessed: 26 January 2017]

[6] Nyirenda SP, Sileshi GW, Belmain SR, Kamanula JF, Mvumi BM, Sola P, et al. Farmers' ethno-ecological knowledge of vegetable pests and pesticidal plant use in Northern Malawi and Eastern Zambia. African Journal of Agricultural Research. 2011;**6**(2):41-49

[7] Amoabeng BW, Gurr GM, Gitau CW, Stevenson PC. Cost: Benefit analysis of botanical insecticide use in cabbage: Implications for smallholder farmers in developing countries. Crop Protection. 2014;**57**:71. DOI: 10.1016/j.cropro.2013.11.019

[8] Dougoud J, Cock MJW, Edgington S, Kuhlmann U. A baseline study using plant-wise information to assess the contribution of extension services to the uptake of augmentative biological control in selected low- to lower-middle-income countries. BioControl. 2018;**63**(1):117. DOI: 10.1007/s10526-017-9823-y

[9] Makaza K, Mabhegedhe M. Smallholder farmers' indigenous knowledge of maize storage pests and pesticidal plant use: The case of wards 9 and 10 in Bikita District, Masvingo Province, Zimbabwe. African Journal of Agricultural Research. 2016;**11**(47):4831

[10] Roy S, Handique G, Muraleedharan N, Dashora K, Roy SM, Mukhopadhyay A, et al. Use of plant extracts for tea pest management in India. Applied Microbiology and Biotechnology. 2016;**100**(11):4831. DOI: 10.1007/s00253-016-7522-8

[11] Stevenson PC, Isman MB, Belmain SR. Pesticidal plants in Africa: A global vision of new biological control products from local uses. Industrial Crops and Products. 2017;**110**:2. DOI: 10.1016/j.indcrop.2017.08.034

[12] Kamatenesi-Mugisha M, Buyungo JP, Egwang P, Vudriko P, Gakunga JN, Deng A, et al. Evaluation of the bio-safety of selected botanical pesticide plants used by subsistence farmers around the Lake Victoria basin. In: Ethnobotany and Health Proceedings of the Cluster Workshop. Uganda: Lake Victoria Research Institute; 2010. pp. 45-57

[13] Belmain SR, Amoah BA, Nyirenda SP, Kamanula JF,

Stevenson PC. Highly variable insect control efficacy of *Tephrosia vogelii* chemotypes. Journal of Agricultural and Food Chemistry. 2012;**60**(40):10055. DOI: 10.1021/jf3032217

[14] Amoabeng BW, Gurr GM, Gitau CW, Nicol H, Munyakazi L, Stevenson PC. Tri-trophic insecticidal effects of African plants against cabbage pests. PLoS One. 2013;**8**(10):e78651. DOI: 10.1371/ journal.pone.0078651

[15] Tembo Y, Mkindi AG, Mkenda PA, Mpumi N, Mwanauta R, Stevenson PC, et al. Pesticidal plant extracts improve yield and reduce insect pests on legume crops without harming beneficial arthropods. Frontiers in Plant Science. 2018;**9**:1425. DOI: 10.3389/fpls.2018.01425

[16] Isman MB. Botanical insecticides: For richer, for poorer. Pest Management Science. 2008;**64**(1):8. DOI: 10.1002/ps.1470

[17] Isman MB. Bridging the gap: Moving botanical insecticides from the laboratory to the farm. Industrial Crops and Products. 2017;**110**:10. DOI: 10.1016/j.indcrop.2017.07.012

[18] Sarasan V, Kite GC, Sileshi GW, Stevenson PC. Applications of phytochemical and in vitro techniques for reducing over-harvesting of medicinal and pesticidal plants and generating income for the rural poor. Plant Cell Reports. 2011;**30**:1163. DOI: 10.1007/ s00299-011-1047-5

[19] El-Wakeil NE. Botanical pesticides and their mode of action. Gesunde Pflanzen. 2013;**65**:125-149. DOI: 10.1007/s10343-013-0308-3

[20] Roobakkumar A, Subramaniam MSR, Babu A, Muraleedharan N. Bioefficacy of certain plant extracts against the red spider mite, *Oligonychus coffeae* (Nietner) (Acarina : Tetranychidae) infesting tea in Tamil Nadu, India. International Journal of Acarology. 2010;**36**(3):255. DOI: 10.1080/01647951003652592

[21] Yang FL, Zhu F, Lei CL. Insecticidal activities of garlic substances against adults of grain moth, *Sitotroga cerealella* (Lepidoptera: Gelechiidae). Insect Science. 2012;**19**(2):205. DOI: 10.1111/j.1744- 7917.2011.01446.x

[22] Zhao NN, Zhang H, Zhang XC, Luan XB, Zhou C, Liu QZ, et al. Evaluation of acute toxicity of essential oil of garlic (*Allium sativum*) and its selected major constituent compounds against overwintering *Cacopsylla chinensis* (Hemiptera: Psyllidae). Journal of Economic Entomology. 2013;**106**(3):1349. DOI: 10.1603/ EC12191

[23] Abdalla MI, Abdelbagi AO, Hammad AMA, Laing MD. Use of volatile oils of garlic to control the cowpea weevil *Callosobruchus maculatus* (Bruchidae: Coleoptera). South African Journal of Plant and Soil. 2017; **34**(3):185

[24] Attia S, Lebdi Grissa K, Ghrabi-Gammar Z, Mailleux AC, Lognay G, Le Goff G, et al. Contrôle de *Tetranychus urticae* par les extraits de plantes en vergers d'agrumes. Faunistic Entomology. 2011;**63**(4):229

[25] Fening KO, Amoabeng BW, Adama I, Mochiah MB, Braimah H, Owusu-Akyaw M, et al. Sustainable management of two key pests of cabbage, *Brassica oleracea* var. capitata L. (Brassicaceae), using homemade extracts from garlic and hot pepper. Organic Agriculture. 2013;**3**(3-4):163. DOI: 10.1007/s13165-014-0058-2

[26] Said F, Inayatullah M, Ahmad S, Khan IA, Saeed-ul-Haq ZM. Comparing the effect of different plant extracts with a chemical insecticide for management of the aphid, Aphis gossypii in sunflower. Pakistan

Journal of Weed Science Research. 2015;**21**(3):359

[27] Baidoo PK, Mochiah MB. Comparing the effectiveness of garlic (*Allium sativum* L.) and hot pepper (*Capsicum frutescens* L.) in the management of the major pests of cabbage *Brassica oleracea* (L.). Sustainable Agriculture Research. 2016;**5**(2):83. DOI: 10.5539/sar.v5n2p83

[28] Degri MM, Sharah HS. Field evaluation of two aqueous plant extracts on water melon *Citrullus lanatus* (Thumb) insect pets in northern Guinea Savannah of Nigeria. International Letters of Natural Sciences. 2014;**9**:59

[29] Kaputa F, Tembo L, Kurangwa W. Efficacy of garlic (*Allium sativum*) and red chilli pepper (Capsicum annum) extracts in the control of red spider mite (*Tetranychus urticae*) in tomatoes (*Lycopersicon esculentum*). Asian Journal of Applied Sciences. 2015;**3**(1): 124

[30] Schmutterer H. Properties and potential of natural pesticides from the neem tree, *Azadirachta indica*. Annual Review of Entomology. 1990;**35**:271-297

[31] Boursier CM, Bosco D, Coulibaly A, Negre M. Are traditional neem extract preparations as efficient as a commercial formulation of azadirachtin A? Crop Protection. 2011;**30**(3):318. DOI: 10.1016/j. cropro.2010.11.022

[32] Shiberu T, Negeri M, Thangavel S. Evaluation of some botanicals and entomopathogenic fungi for the control of onion thrips (*Thrips tabaci* L.) in West Showa, Ethiopia. Journal of Plant Pathology & Microbiology. 2012;**04**:01. DOI: 10.4172/2157-7471. 1000161

[33] Ibrahim A, Demisse G. Evaluation of some botanicals against termites' damage on hot pepper at Bako, Western Ethiopia. International Journal of Agricultural Policy and Research. 2013;**1**(2):48

[34] Aziz MA, Ahmad M, Nasir MF. Efficacy of different neem (Azadirachta indica) products in comparison with Imidacloprid against English grain aphid (*Sitobion avenae*) on wheat. International Journal of Agriculture and Biology. 2013;**15**(2):279

[35] Degri M, Mailafiya D, Wabekwa J. Efficacy of aqueous leaf extracts and synthetic insecticide on pod-sucking bugs infestation of cowpea (*Vigna unguiculata* (L.) Walp) in the Guinea Savanna Region of Nigeria. Advances in Entomology. 2013;**01**(02):10. DOI: 10.4236/ae.2013.12003

[36] Kumar MM, Kumar S, Prasad CS, Kumar P. Management of gram pod borer, *Helicoverpa armigera* (Hubner) in chickpea with botanical and chemical insecticide. Journal of Experimental Zoology, India. 2015;**18**(2):741

[37] Patil RS, Nandihalli BS. Efficacy of promising botanicals against red spider mite on brinjal. Karnataka Journal of Agricultural Sciences. 2009;**22**(3):690

[38] Ilesanmi JO, Gungula DT. Preservation of cowpea (*Vigna unguiculata* (L.) Walp) grains against cowpea bruchids (*Callosobruchus maculatus*) using neem and moringa seed oils. International Journal of Agronomy. 2010;**2**:1. DOI: 10.1155/2010/235280

[39] Ileke KD, Oni MO. Toxicity of some plant powders to maize weevil, *Sitophilus zea mais* (Motschulsky) [Coleoptera: Curculiondae] on stored wheat grains (*Triticum aestivum*). African Journal of Agricultural Research. 2011;**6**(13):3043. DOI: 10.5897/AJAR11.622

[40] Kemabonta KA, Falodu BB. Bio-efficacy of three plant products as post-harvest grain protectants against *Sitophilus oryzae* Linnaeus (Coleoptera:

Curculionidae) on stored wheat (*Triticum aestivum*). International Journal of Natural Sciences. 2013;**4**(2):259

[41] Paul UV, Lossini JS, Edwards PJ, Hilbeck A. Effectiveness of products from four locally grown plants for the management of *Acanthoscelides obtectus* (Say) and *Zabrotes subfasciatus* (Boheman) (both Coleoptera: Bruchidae) in stored beans under laboratory and farm conditions in Northern Tanzania. Journal of Stored Products Research. 2009;**45**(2):97. DOI: 10.1016/j.jspr.2008.09.006

[42] Wakeil N, Gaafar N, Sallam A, Volkmar C. Side effects of insecticides on natural enemies and possibility of their integration in plant protection strategies. In: Insecticides—Development of Safer and More Effective Technologies. London: InTech; 2013. pp. 3-56

[43] Sithisut D, Fields PG, Chandrapathya A. Contact toxicity, feeding reduction and repellency of essential oils from three plants from the ginger family (*Zingiberaceae*) and their major components against *Sitophilus zeamais* and *Tribolium castaneum*. Journal of Stored Product. 2011;**104**:1445-1454

[44] Regnault-Roger C, Vincent C, Arnason JT. Essential oils in insect control: Low-risk products in a high-stakes world. Annual Review of Entomology. 2012;**57**:405-424. DOI: 10.1146/annurev-ento-120710-100554

[45] Wafaa MH, Rowida S, Baeshen, Hussein AH, Ahl S-A. Botanical insecticide as simple extractives for pest control. Cogent Biology. 2017;**3**(1): 1404274

[46] Enan EE. Molecular and pharmacological analysis of an octopamine receptor from american cockroach and fruit fly in response to plant essential oils. Archives of Insect Biochemistry and Physiology. 2005;**59**:161-171

[47] Verma N, Tripathi AK, Prajapati V, Bahl JR, Bansal RP, Khanuja SPS, et al. Toxicity of essential oil from Lippia alba towards stored grain insects. Journal of Medicinal & Aromatic Plants. 2001;**22/4A and 23/1A**:117-119

[48] Ahmed N, Alam M, Saeed M, Ullah H, Iqbal T, Al-Mutairi KA, Shahjeer K, Ullah R, Ahmed S, Ahmed NA, Khater HF. Botanical Insecticides Are a Non-Toxic Alternative to Conventional Pesticides in the Control of Insects and Pests. global Decline of Insects 2021 Oct 25. IntechOpen.

[49] Tripathi AK, Prajapati V, Khanuja SPS, Kumar S. Effect of d-limonene on three stored product beetles. Journal of Economic Entomology. 2003a;**96**:990-995

[50] Tripathi AK, Prajapati V, Kumar S. Bioactivities of l-carvone, d-carvone and dihydrocarvone towards three stored product beetles. Journal of Economic Entomology. 2003;**96**: 1594-1601

[51] Tripathi AK, Prajapati V, Verma N, Bahl JR, Bansal RP, Khanuja SPS, et al. Bioactivities of the leaf essential oil of Curcuma longa (Var. CH-66) on three species of stored product beetles. Journal of Economic Entomology. 2002;**95**:183-189

[52] Arun KT, Shikha U, Mantu B, Bhattacharya PR. A review on prospects of essential oils as biopesticide in insect-pest management. JPP. 2009;**1**(5):052-063

[53] FAO, WHO. Manual on Development and Use of FAO and WHO Specifications for Pesticides, First Edition-Third Revision. FAO Plant Production and Protection Paper. Vol. 228. Rome: World Health Organization; Food and Agriculture Organization of the United Nations; 2016

Chapter 8

# Fenitothion Degradation by *Aspergillus parasiticus*

*Thenepalli Sudha Rani and Potireddy Suvarna Latha Devi*

## Abstract

India is a predominantly agriculture-based country with a population of 1.27 billion, according to FAO the population has reached to 1.66 billion in between 2007 and 2050. Tense because of overgrowing population the yield of crops were increased by applying various insecticides for controlling (insects, pests). Globally, an appraise 1 to 2.5 million tons of effective insecticide additives go on applied each year, especially in agriculture. Fenitrothion is an organophosphate insecticide employed to destroy pests, insects particularly in Paddy fields and it is an acetylcholinesterase inhibitor, neurotoxicant and the toxic metabolites in the environment is remain for longer periods, so it is necessary to degrade the fenitrothion by biodegradation. The fungi *Aspergillus parasiticus* were screened from paddy fields and Molecular characterized it by 26S rDNA gene sequencing, the fungi breaks the insecticide within 24 h of incubation in PDB. The course of the degradation process was studied using FTIR and HPLC.

**Keywords:** FAO, FTIR, HPLC, 26S rdna, acetylcholinesterase inhibitor

## 1. Introduction

Extensive dimensions of insecticides make employment toward agriculture everywhere in the universe [1]. Organophosphates (OPs) remain a class of insecticides, certain of which are extremely toxic. Organophosphorus composite poisoning is a global health obstacle among nearby 3 million poisonings furthermore 200 000 deaths periodically [2]. The primary organophosphorus insecticide, tetraethyl pyrophosphate, did originate furthermore employed near 1937. They were among the numerous extensively applied insecticides available. Organophosphates (also known as phosphate esters, or OPEs) are a group regarding organophosphorus compounds besides the global edifice $O{=}P(OR)_3$, a prime phosphate particle including alkyl or aromatic substituents.

Most utmost of the organophosphorus insecticides enhance relevent universal composition, comprising 3 phosphoester linkages, plus hydrolysis concerning one of the phosphoester bonds dramatically diminishes the toxicity regarding comic insecticides by eliminating their acetylcholinesterase–inactivating properties [3] fenitrothion [O, O-dimethyl O- (3-methyl-4-nitrophenyl) phosphorothioate] stands one of the various broadly utilized broad-spectrum organophosphate insecticide, an acaricide is exercised to slaughter pests like piercing, crunch, and suctorial insect pests (bugs of wheat, beetles of flour, grain, stem borers of rice, Weevils of grain) usually acts on rice, cereals, grasp, greens, further applied as a mosquito, sail, cockroach repellents, sprays as fields and society curriculum, and can do grasped through each route, including inhaling, ingestion, plus dermal intake, comic toxicological effect as concerns comic insecticide fenitrothion is about entirely due to comic repression of acetylcholinesterase in the nervous system, emerging against respiratory, myocardial and neuromuscular transmission impairment, and comic toxic metabolites in the environment remain for more sustained periods, because of this inference fenitrothion is degraded by employing microorganisms.

The extreme degradation product of fenitrothion is 3-methyl-4-nitrophenol and induces extensive corruption in soils and the aquatic environment.

Fenitrothion degradation proceeds through hydrolysis & photolysis under sunlight (or) UVR, microflora further impersonates a very important role in degradation, fenitrothion in water is stable when microorganisms, sunlight is not an available form, in soil mainline of degradation is the biodegradation [4].

In the degradation of fenitrothion the biological spp. namely *Anthrobacter aurescens, Burkholderia, Rhizobium, Flavobacterium, Chlorella, Pseudomonas* operate vital activity.

Genes of particular are abode culpable for the degradation of insecticides, several practicable organisms able of cleavage diverse sort of organophosphates has been screened and characterized from various slots. The vastly optimistic way for the degradation of organophosphate insecticides are enzymatic mechanisms, of extracellular, hydrolase of organophosphorus (OPH) has being a classic enzyme with capable to resolve a vast array of organophosphorus pesticide and chemical combat determinants.

Organophosphate hydrolase along with MPD, OPD, MPH, etc. regard to organophosphorus hydrolase group. Organophosphorus hydrolase has the highest exercise along with wide-ranging for the abrupt withdrawal out of organophosphates. Along with this parathion hydrolase, paraoxonase, esterase, phosphotriesterase, and diisopropyl fluorophosphatase [5] also display the pivotal part in fenitrothion degradation. Pakala [6] identified the bacterial species namely *Serratia*, which is responsible for the deterioration of Parathion by Parathion hydrolase and also reducing the nitro group by nitroreductase.

Bioremediation is the process of applying biological systems of diminution in regard to contaminants of aquatic, sublunary or from the wind. The major biological systems applied for Bioremediation are naturally or premeditatively microorganisms and plants. The most frequently applying method for Biodegradation is bioremediation with microorganisms. The present universal proceedings of bioremediation that comprise bioengineering tense potential of innate microbes to clear up comic habitat are compelling different to prevalent rendition methods [7].

Fungi, bacteria acts as the cheaper, excel environment friendlier option in deterioration referring to insecticides of organophosphates, biological process form different metabolites in fenitrothion degradation. Biodegradation was effective in the treatment of this pollution in a eco-friendly manner.

In the present study, a novel fungal *Aspergillus Parasiticus* capable as concerns deteriorating not justly fenitrothion but also 3-methyl-4-nitrophenol was isolated. Biodegradation of fenitrothion in Czepak-dox medium was studied. Tense research directs toward elucidating per probable employment of an isolated fungal strain toward remediation as concerns comic fenitrothion-contaminated environment.

## 2. Materials and methods

### 2.1 Chemicals

Analytical grade fenitrothion (50% ec) were purchased from Shijiazhuang Awiner Biotech Co., Ltd., China and were employed as standard. Technical grade fenitrothion a 20% emulsifiable concentrate used in this study were obtained from Chennai local market, India. All additional reagents applied in this study were of high purity and analytical grade.

### 2.2 Soil sample

Paddy soils were taken away from the agriculture field regarding Pakala, Chittoor District, Andra Pradesh, with a sustain cultivation exercise as well as Thirty years. Exterior clay of 0–15 cm was levelheaded, stored currently in elastic pouches, transferred these particulates directed toward the lab [8]. The paddy clay was drained at room temperature, restrained the wateriness contentment by about 20% (W/W), besides, the paddy clay is transpired over a bowl-shaped sieve alongside a 2 mm net, physicochemical parameters of the test clay was assayed, detailed physicochemical parameters of the soil are presented in **Table 1**.

### 2.3 Isolation of fungi-enrichment technique

The clay samples of 50 g were allocated in various Erlenmeyer flasks and samples were further enriched with amendment of different (10, 25, 50, 75,100 ppm) concentrations with fenitrothion respectively to provide a terminal quantity about 100 ppm, agitated the flasks forcefully being homogeneous mingle of insecticide, incubate it at 27C ± 2°C up to 3 weeks, wateriness contentment was maintained with the addition of distilled water twofold by 1-week interim [8]. The media of stock culture were processed over transpose 5 g of paddy clay to Potato dextrose broth from enriched clay samples of pH -7 [9] with ingredients Potatoes, infusion-200.0, Dextrose-20.000, Agar - 15.000, pH (at 25°C)- 5.6 ± 0.2 Gms/lt, without agar for preparation of broth.

### 2.4 Screening of Fenitrothion degrading fungi

10 ml of the stock cultures abide transmitted into the range of 100 ml Erlenmeyer flask consist of fresh 50-ml broth of Potato dextrose, subject it to the incubator belongs to shaking by maintaining the speed of rpm 250 at 27C ± 2°C. Further, the culture of 1 ml is transferred to clean Erlenmeyer flasks containing fresh broth with (10, 25, 50, 75,100 ppm) of insecticide and maintained at 27C ± 2°C with shaking at 250 rpm for 1 week. The steps were repeated up to 6 transfers. Following 6 transmittals, one loop of inoculates abide inoculated over agar of Potato dextrose plates, stored by 27C ± 2°C for 24 - 48 h [8].

| Parameters | Soil |
|---|---|
| Sand (%) | 50% |
| Silt (%) | 20% |
| Clay (%) | 27% |
| Organic matter (%) | 0.8% |
| Texture | Black loam-sandy clay |
| pH | 7.8 |
| Maximum water-holding capacity (ml/g) | 0.226 |

**Table 1.**
*Physicochemical parameters of the test soil.*

### 2.5 Enrichment procedure for isolation of potential fungi strain

Fungal isolates were carried out in Czepak-dox broth according to the methods of [10]. Fungi isolate degrading fenitrothion were obtained by enrichment culture in the Czepak-dox agar media containing Sodium nitrate, 2.0 gm $L^{-1}$, Sucrose, 30.0 gm $L^{-1}$, Magnesium sulphate, 0.5 gm $L^{-1}$, Dipotassium Phosphate, 1.0 gm $L^{-1}$, Ferrous sulphate, 0.01 gm $L^{-1}$, Potassium Chloride, 0.5 gm $L^{-1}$, Agar, 15.0 gm $L^{-1}$, pH 7.3 ± 0.3 at 25°C by successively greater fenitrothion convergence (200, 300, 400, 500 ppm) by maintaining the controls (without inoculation of fungi). For this different ppm concentration of fenitrothion (**Figure 1**) were prepared by solubilizing the fenitrothion in acetone. By using Czepak-dox broth the degradation of insecticide is also checked in liquid media [7, 8, 11].

### 2.6 Growth studies of the potential isolate

Growth curves of fungi isolate were determined in PDB with fenitrothion and without fenitrothion as control, A culture of aliquant is taken out by constant interim as for 0, 5, 10, 15, 20, 24 hours. Absorbance was measured at 600 nm [12].

### 2.7 Parameters of optimization

To check shaking & static consequence of insecticide degradation, flasks containing 50 ml PDB amended with fenitrothion insecticide and fungi culture were inoculated and incubated at 37°C in static condition and another set is subject to the shaker of orbital up to 24 hours by 120 rpm. Insecticide samples are introverted by systematic span interregnum & exposed to degradation assay.

### 2.8 Utilization of phosphate by fungi

According to the literature of [7, 8, 11, 13]. The fungi utilize phosphorus from [Organo Phosphate Insecticide] as the major source for their growth- Phosphatase activity. Czepak-dox agar medium with, without out Phosphorus & dispersed in conical flaskets of 100 ml & sterilized with autoclave through standard manner, after sterilization various concentrations related to Fenitrothion of 50% EC as 10, 20, 50, 100 ppm is added as a phosphorus source. Two agar plates were kept as Control - Czepak-dox agar medium with Phosphorus (without Fenitrothion). The Isolate namely *Aspergillus parasitcus* abide cleft in distinction to earnestly thriving culture on PDA & positioned on comic centre as concerns specific Petri

**Figure 1.**
*Different ppm concentrations of fenitrothion.*

dishes encompass a various congregation as for Fenitrothion. The effect of growth and utilization of Phosphate by *Aspergillus parasitcus* by virtue of culture medium belongs to liquid prior to Czapak Dox, be accomplished with 2 calibrates albeit with & without Fenitrothion emendation.

### 2.9 Taxonomic identification of the fungi strain

Genomic DNA isolation purification is carried out by carried out by utilizing fungal-peculiar 26 s rDNA sequencing of gene molecular characterization [14] was identified. Further, strains were amplified by PCR and then confirmed by molecular-based 26 s rDNA partial sequencing accomplished at National Collection of Industrial Microorganisms (NCIM) CSIR-NCL, Pune. Virtually intact term 26 s rDNA abide ampliate over PCR upon ITS1, ITS4. By using the universal primer this reaction was carried out. The sequel of primer abide follows in the process of 5'TCCGTAGGTGAACCTGCGG3'- ITS1 5'TCCTCCGCTTATTGATATGC3'- ITS4, the polymerase chain reaction were carried out by Initiatory denaturation-94°C-5 min, Denaturation-94°C-30 sec, Annealing 56°C-30 sec, Elongation 72°C-30 sec, Eventual expansion 72°C-10 min up to 35 cycles.

### 2.10 Insecticide residues-extraction & exploration

Tense flasks of culture Test Sample be possessed & percolate via the filter paper of Whatman No.1, elicitation regarding Fenitrothion amid extract of culture filtrate [15] abide accomplished later. The filtrate abides embrace as the funnel of detached facing that saline water of twenty ml about the percentage of 2 was put on. Subsequently, hexane of 40 ml, 20 ml of ethyl acetate was put on, vibrate the funnel aggressively & and concede on the point of base up to ten minutes thus 2-phases simultaneously aqueous, organic phase comes into detached, tense pace be redone thrives on redeeming comic insecticide. Finishing, funnel abide permit on sit up to fifteen minutes in-favor-of entire detachment regarding phases. Tense upper layer (organic phase) comprises fenitrothion be separated & the samples containing the residues of fenitrothion were subject for chromatographic procedures.

### 2.11 FTIR-Fourier-transform infrared spectroscopy-interpretation

Deterioration products as for the fenitrothion ensue monitored on FTIR – utilized as investing modifications with it apparent functional categories such abide intricate with its comic degradation about fenitrothion. The sample & control abide torrid & assorted by Potassium bromide (1:20; 0.02 grams as regards to sample accompanied by KBr with finishing net of 0.4 grams) Premise the samples, desorb it by 60°C & press down for pellets of IR-transparent. Tense absorbancy spectra regarding samples are chronicled by utilizing (FT-IR-NICOLET IS10). The scanning rate as concerns 500–3000 $cm^{-1}$ is applied for taking the spectra. Tense FT-IR is initially measured peculiar background scanning along with control as clear Potassium bromide & afterwards, the sample regarding analysis be scanned tense Fourier-transform infrared spectrum about comic non-deteriorated control be finishing deduct out of possession of comic spectra about deteriorated insecticide [16]. The positions of stretching & band, bending be espied & collate along with allusion compounds. With wave quantity group the band posture is conferred ($cm^{-1}$ reciprocal centimeters). Tense band ferocity manifested as (T) transmittance. According to comic band positions, the presence of functional groups was counted.

### 2.12 HPLC - high-performance liquid chromatography

Tense deteriorated compounds abide determined at high-performance liquid (HPLC-1200 series) chromatography. Decolourized residue was dissolved in acetonitrile was inject into the column using mobile phase like acetonitrile-water. HPLC be carried out to separate individual compounds of intermediates, for separation of sole products concerning intermediary that were identified by utilizing of UV–Vis detector reverse phase column be applied. Tense acetonitrile-water in the ratio of 1:1 was used with the rate of movement 0.5 ml/minute. Tense eluates are monitored by 254 nm wavelength using isocratic elution [17].

## 3. Results and discussion

In the current study, we practised selective enrichment methods to isolate fenitrothion deteriorating fungi of the paddy field and 5 distinct strains was obtained, among which *Aspergillus parasitcus* was chosen for analysis because of potential degradation of fenitrothion. The fungi utilize fenitrothion as phosphate source.

Soil sample collected from paddy field was enriched with fenitrothion to isolate the fenitrothion degrading fungi. From this enrichment culture, among 5 distinct strains were isolated on Potato dextrose medium containing fenitrothion. Czapek Dox Agar plate applied to screen the isolates for potential tolerance to fenitrothion.

BLAST result of the 26 s rDNA gene sequence of fungi isolate exhibited 99% similarity to that of the 26 s rDNA gene of *Aspergillus parasitcus* (GenBank accession no. MH714745). (**Figure 2**) indicates growth kinetics of *Aspergillus parasitcus*, the metabolism of fenitrothion by *Aspergillus parasitcus* was indicated by a visible increase in mycelia mass with time, the growth curve pattern was studied by growing the organisms in the presence of insecticide and comparing it with the control (without insecticide). The growth pattern of *Aspergillus parasitcus* was significantly different from the control and the lag phase delayed up to 12 hours in comic residence as concerns both isolates while in comparison toward control. Tense maximum progress was observed after 21 hours in *Aspergillus parasitcus*. The number of cells decreased as fenitrothion degradation progressed in time. Tense cells eventually are old, lyse & comic enzyme of extracellular interacts with insecticide to reduce the toxicity.

The degradation efficiency of fenitrothion insecticide was studied by static and shaking conditions at various time intervals. The resolute of degradation were identified through an increase at the flasks to be retained in a condition of static (90%) and comic activity of degradation was reduced beneath the condition of

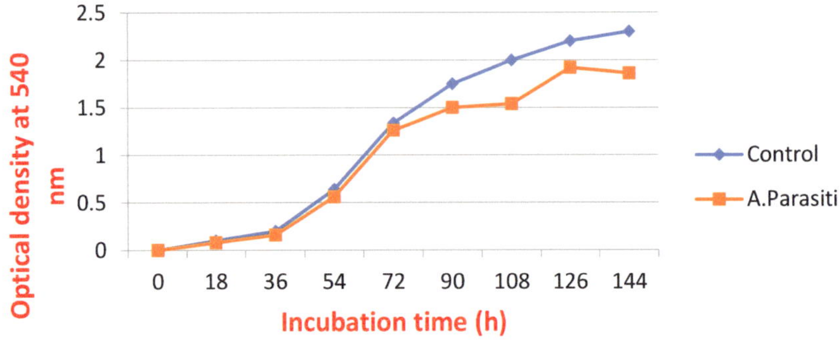


**Figure 2.**
*Growth curve of* Aspergillus parasitcus.

shaking (30%) (**Figure 3**). Beneath the conditions of shaking, oxygen presence divests hydrolase enzyme so the degradation process decreased, whereas under static conditions the activation of enzyme degrades the fenitrothion.

Fungi Utilize Fenitrothion as Phosphate Source when compared to control. Control with lack of fenitrothion, the 2 fungi with fenitrothion shows similar growth rates, which intimates phosphate is the major source for the growth of 2 isolated fungi (in solid, liquid media) namely *Aspergillus parasitcus* intimated in the (**Figures 4** and 5) and (**Figures 6** and 7).

### 3.1 In liquid medium

The spectrum of FTIR *Aspergillus parasiticus* is analyzed between the scan ranges (500–3500 $cm^{-1}$), The FTIR spectrum obtained from the control (**Figure 8**)

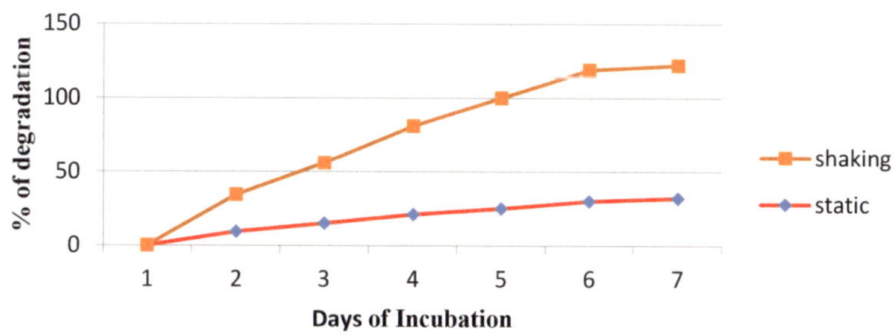


**Figure 3.**
*Effect of stationary & shaking situation on the degradation of insecticide.*

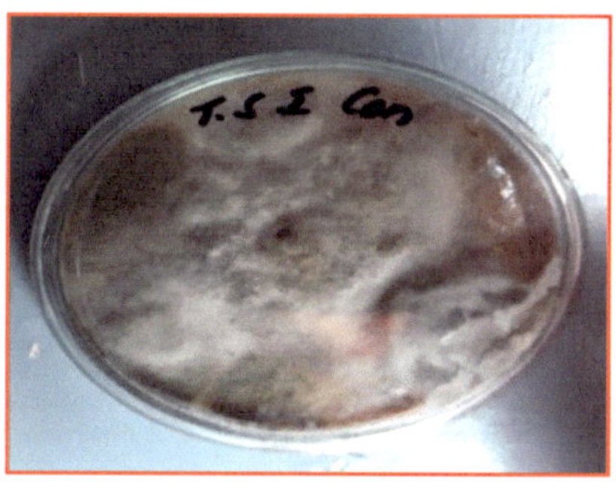

**Figure 4**
Control (without fenitrothion).

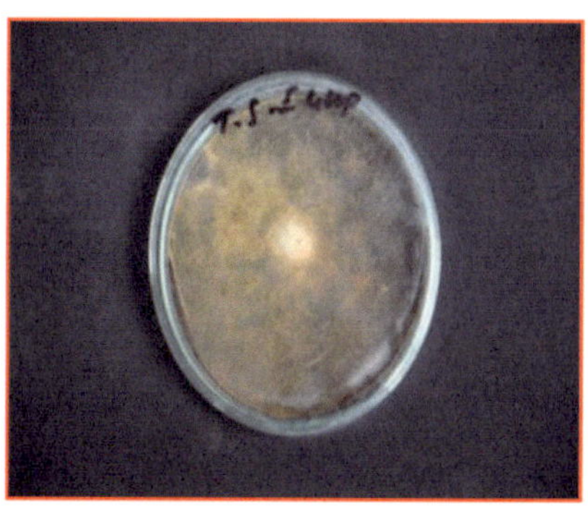

**Figure 5.**
A. parasitcus *(with fenitrothion).*

**Figure 6.**
*Control (without fenitrothion).*

**Figure 7.**
A. parasitcus *(with fenitrothion).*

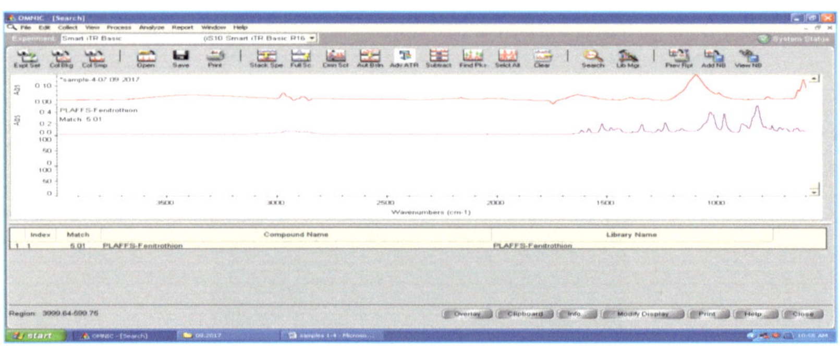

**Figure 8.**
*FTIR spectrum of A. parasiticus (FI-I) (red-Control, Purple- FI-1).*

[Peaks of red color intimates control] displayed a peak at 2950 $cm^{-1}$ 2850 $cm^{-1}$ indicating stretching and strong vibration of the C-H bond of alkane (**Table 2**). A Peak at 2150 indicating bending and medium-weak vibration of C-H bond of alkane. A Peak on 1700 $cm^{-1}$ & 1600 $cm^{-1}$ exhibit C=O lengthen & strong vibration of the carbonyl group. Peaks on 1350 $cm^{-1}$ & 1300 $cm^{-1}$ lengthen vibration

& medium-weak of alkyl Halide compounds. Peaks at 1250 $cm^{-1}$ exhibit C-N lengthen as regards amine compounds. Peaks on 1080 & 1030 $cm^{-1}$ exhibit C-O C-O lengthen as for strong compounds of ether. Peaks at 980 $cm^{-1}$ & 750 $cm^{-1}$ showed = C-H alkene compounds. Peaks by 620 $cm^{-1}$ and 610 $cm^{-1}$ & 600 $cm^{-1}$ exhibit Stretch & Strong vibration of alkyl halide respectively.

The FTIR spectrum of the products formed after degradation in *Aspergillus parasiticus* isolate (**Figure 8**) [Peaks of purple color intimates] array a peak on 2910 $cm^{-1}$ & 2890 $cm^{-1}$ showed C-H indicating lengthen and strong vibration of alkane compounds. The peak on 1600 $cm^{-1}$ exhibit N-H indicating lengthen and strong quaking of amine compounds. Peak on 1590 $cm^{-1}$ exhibit C=C indicating lengthen and strong quaking aromatic compounds. Tense peak at 1500 $cm^{-1}$ exhibit N-O indicating lengthen & strong vibration intimates nitro groups. Peaks on 1490 $cm^{-1}$ & 1400 $cm^{-1}$ showed C=C lengthen of aromatic compounds. Peaks on 1350 $cm^{-1}$ showed C-N stretching concerning strong aromatic amines. Peaks on 1330 $cm^{-1}$ showed C-N bent concerning strong aromatic amines. Peaks on 1230 $cm^{-1}$ showed O-H lengthen concerning strong alcohol compounds. Peaks at 1220 $cm^{-1}$ and 1010 $cm^{-1}$ showed = C-H alkyl halide compounds. Peaks on 980 $cm^{-1}$ showed O-H Stretch and Strong vibration of carboxylic acids. Peaks at 790 $cm^{-1}$ showed = C-H bending and Strong quaking concerning an alkene. Peaks on 670 $cm^{-1}$, 660 $cm^{-1}$, 640 $cm^{-1}$, 630 $cm^{-1}$ showed O-H Stretch, bent and Strong vibration of alkyl halide Peaks at 600 $cm^{-1}$ & 590 $cm^{-1}$ exhibit C=C lengthen & Strong vibration of aromatic compounds respectively (**Table 2**).

| **Functional group** | | **Type of Vibration** | **Characteristic Absorptions ($cm^{-1}$)** | **Intensity** |
|---|---|---|---|---|
| **Control** | | | | |
| Alkane | C-H | Stretch | 2950 | Strong |
| Alkane | C-H | Stretch | 2850 | Strong |
| Alkane | C-H | Bending | 2150 | Medim-weak |
| Carbonyl | C=O | Stretch | 1700 | Strong |
| Carbonyl | C=O | Stretch | 1600 | Strong |
| Alkyl Halide | C-H | Stretch | 1350 | Medim-weak |
| Alkyl Halide | C-H | Stretch | 1300 | Medim-weak |
| Amine | C-N | Stretch | 1250 | Strong |
| Ether | C-O | Stretch | 1080 | Strong |
| Ether | C-O | Stretch | 1030 | Strong |
| Alkene | =C-H | Stretch | 980 | Strong |
| Alkene | =C-H | Stretch | 750 | Medim-weak |
| Alkyl Halide | C-H | Stretch | 620 | Strong |
| Alkyl Halide | C-H | Stretch | 600 | Strong |
| Alkyl Halide | C-H | Stretch | 610 | Strong |
| *Aspergillus parasiticus* **(FI-1)** | | | | |
| Alkane | C-H | Stretch | 2910 | Strong |
| Alkane | C-H | Stretch | 2890 | Strong |
| Amine | N-H | Stretch | 1600 | Strong |
| Aromatic | C=C | Stretch | 1590 | Strong |

| Functional group | | Type of Vibration | Characteristic Absorptions (cm$^{-1}$) | Intensity |
|---|---|---|---|---|
| Nitro | N-O | Stretch | 1500 | Strong |
| Aromatic | C=C | Stretch | 1490 | Strong |
| Aromatic | C=C | Stretch | 1400 | Strong |
| Aromatic amines | C-N | Stretch | 1350 | Weak |
| Aromatic amines | C-N | Bended | 1330 | Strong |
| Alcohol | O-H | Bended | 1230 | Weak |
| Alkyl Halide | C-H | Stretch | 1220 | Strong |
| Alkyl Halide | C-H | Stretch | 1010 | Strong |
| Carboxylic acids | O-H | Stretch | 980 | Strong |
| Alkene | =C-H | Bending | 790 | Strong |
| Alkyl Halide | C-H | Stretch | 670 | Strong |
| Alkyl Halide | C-H | Stretch | 660 | Strong |
| Alkyl Halide | C-H | Bending | 640 | Weak |
| Alkyl Halide | C-H | Stretch | 630 | Strong |
| Aromatic | C=C | Stretch | 620 | Strong |
| Aromatic | C=C | Stretch | 600 | Strong |
| Aromatic | C=C | Stretch | 590 | Strong |

**Table 2.**
*FTIR compounds from fenitrothion degrading from* Aspergillus parasiticus.

The insecticides were determined on collation based on comic retention time by samples with a comic standard. The HPLC elution profile of fenitrothion (control) showed prominent peaks at retention time of 10.652 minutes (**Figure 9**). The samples at 3–4 days of the interval beyond be a notable decline by the magnitude appropriate to peak on retention time 2.489,1.950, 1.275,1.209 (**Figure 10**) & The samples at 7–8 days of the interval a notable decline by magnitude appropriate to peak on retention time 1.930, 1.231 (**Figure 11**) in the degraded sample *Aspergillus parasiticus* confirming the degradation of fenitrothion. Various peaks do too espy

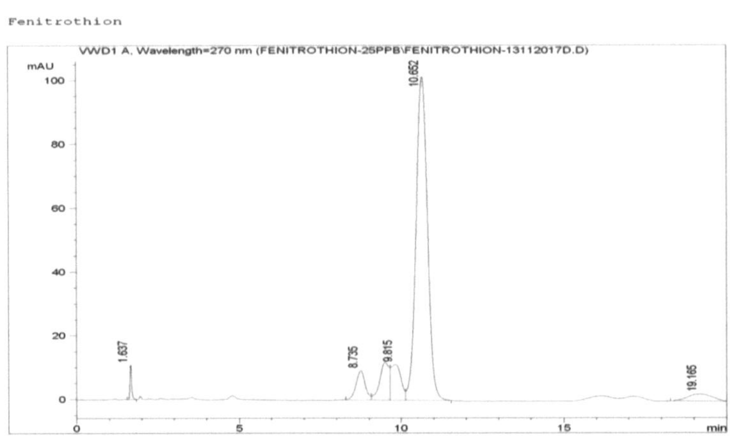


**Figure 9.**
*HPLC chromatogram of fenitrothion (control).*

| Peak # | R.T [min] | Type | Width [min] | Area | Area% |
|---|---|---|---|---|---|
| 1 | 1.637 | VB | 0.057 | 40.616 | 1.221 |
| 2 | 8.735 | BV | 0.302 | 187.882 | 5.647 |
| 3 | 9.485 | VV | 0.301 | 245.940 | 7.392 |
| 4 | 9.815 | VV | 0.345 | 255.086 | 7.667 |
| 5 | 10.652 | VB | 0.383 | 2480.939 | 74.572 |
| 6 | 19.165 | BBA | 0.629 | 116.432 | 3.500 |

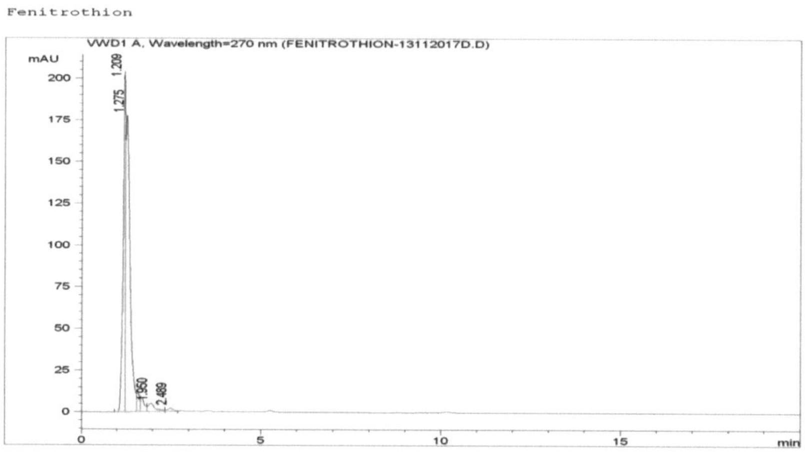


**Figure 10.**
*HPLC analysis of fenitrothion degradation by* Aspergillus parasiticus.

| Peak # | R.T [min] | Type | Width [min] | Area | Area% |
|---|---|---|---|---|---|
| 1 | 1.209 | BV | 0.062 | 226.236 | 91.211 |
| 2 | 1.275 | VV | 0.082 | 1032.986 | 25.875 |
| 3 | 1.950 | VB | 0.211 | 24.360 | 1.809 |
| 4 | 2.489 | BB | 0.419 | 1856.630 | 53.471 |

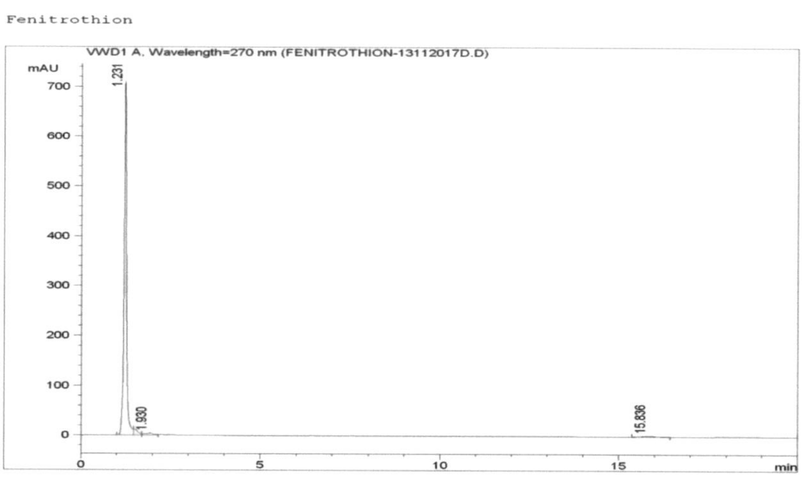


**Figure 11.**
*HPLC analysis of fenitrothion degradation by* Aspergillus parasiticus.

| Peak # | R.T [min] | Type | Width [min] | Area | Area% |
|---|---|---|---|---|---|
| 1 | 1.231 | BV | 0.072 | 235.266 | 94.562 |
| 2 | 1.524 | VV | 0.110 | 1032.761 | 2.556 |
| 3 | 1.930 | VV | 0.204 | 63.226 | 1.705 |
| 4 | 15.836 | BB | 0.360 | 43.632 | 1.177 |

by comic chromatogram as regards the degraded sample illustrate comic proffering as concerns metabolites at comic isolates. The significant absence concerning comic peaks recognized by comic insecticide (control) sample & tense presence as concerns strange peaks at comic degraded metabolites upon strange retention times ramparts comic biotransformation as regards parent insecticide toward molecules.

## 4. Conclusions

Fenitrothion organophosphate insecticide was selected for the present study. It is well known to be poisonous, carcinogenic, mutagenic and pollutant in nature because it is an acetylcholine esterase inhibitor, and inhibit different metabolic activities, and also highly toxic to all living ecosystem.

The degradation ability is also observed under static/ shaking conditions and was measured by spectrophotometric method. Degradation of fenitrothion insecticide was more efficient in static condition than at shaking condition with 7 days of incubation. The static condition, transmit as regards oxygen abide finite toward comic surface of the broth & comic cell cultures do utmost probably residuum of comic flasks & get briskly drained oxygen and enhance the degradation, fungi produce an enzyme which helps to break down the organic compounds in wastewater.

In this study the fungi utilize the phosphate as the major nutrient for their growth which is tested by Czapek Dox media which were prepared with and without fenitrothion, without fenitrothion were coded as control, the plates which contain fenitrothion should be inoculated as the two selected fungi spp. as the same way the control is also inoculated with fungi, and kept for incubation up to 7 days, along with control, the plates which contain fenitrothion also shows the same growth, here while preparation of the Czapek Dox media for control all nutrients namely sodium nitrate, sucrose, magnesium sulphate, dipotassium phosphate, ferrous sulphate, potassium chloride was added, here dipotassium phosphate serves as a phosphate source for the growth of 2 fungi, while in the other plates dipotassium phosphate is not added, instead of this fenitrothion is added, fungi for their growth it utilizes the fenitrothion as a phosphate source. Different temperature and pH and different time intervals also influence the growth of fungi which is a helpful factor to know the detailed conditions of the selected fungi.

Differences in the FTIR spectrum of fenitrothion and metabolites indicated that the insecticide molecule degraded into different metabolites. In FTIR analysis, control (insecticide) had several peaks. The difference in the FTIR spectrum of fenitrothion and metabolites indicated that the insecticide molecule degraded into different metabolites by *Aspergillus parasiticus*. The presence as concerns latest peaks in comic insecticide and nonappearance appropriate to the above peak representing the catalyzed cleavage of fenitrothion.

In the FTIR spectrum, exhibit an important modification over the position as concerns a peak, while correlated toward comic control insecticide span in both fungi isolates of *Aspergillus parasiticus*. Significant disappearance of the peaks develops over comic insecticide sampler & comic emergence as concerns fresh peaks by comic degraded samplers beside fresh retention times rampart comic biotransformation as concerns fenitrothion toward fresh compounds.

The HPLC chromatogram of fenitrothion showed prominent peaks at retention time 10.652 intimates the control, the reduction chic ferocity as concerns comic peak by *Aspergillus parasiticus* retention time was 1.275, 1.209, 1.950, 2.489 and 1.231, 1.930, 15.836 respectively. The study exhibit comic appearance as regards peaks amidst the vanishing of the peaks of fenitrothion confirming the insecticide degradation by metabolites. The results supported by the emergence as concerns of the latest peaks over comic deteriorated compounds concoct later degradation, due to the production of different intermediate metabolites.

Our study revealed that the fungi isolate exhibit an increased level of degradation at 300 ppm concentration. Fenitrothion at a concentration of 100 to 400 ppm observed an increase in degradation with the increase in insecticide concentration. At lower ppm concentrations 75, 100 the degradation rates were increased rapidly, but the captivation as regards insecticide be boost amid 200–300 ppm comic deterioration rates were very slowly at starting days of incubation but on prolonged incubation, up to 14 days the degradation rates were increased, intimates that rapid increase in ppm concentration will slow the growth of organisms. When the concentration of ppm up to 400 ppm and 7 to 14 days there is no growth, but on prolonged incubation up to 20 days the degradation rates were increased slowly.

This intimates that at higher concentrations of fenitrothion up to 600, 1000 ppm also shows degradation from slow level to a higher level. Here at the time insecticide concentration was high, the isolates showed less capability.

## Author details

Thenepalli Sudha Rani* and Potireddy Suvarna Latha Devi
Department of Applied Microbiology, Sri Padmavati Mahila Visvavidyalayam, Tirupati, A.P., India

*Address all correspondence to: sudhapakala84@gmail.com

## References

[1] Shukla G, Kumar A, Bhanti M, Joshep PE, Taneja A. Organochlorine pesticide contamination of ground water in the city of Hyderabad. Environment International. 2006;**32**(2):244-247. DOI: 10.1016/j.envint.2005.08.027

[2] Karalliedde L, Senanayake N. Organophosphorus insecticide poisoning. British Journal of Anaesthesia. 1989;**63**(6):736-750. DOI: 10.1093/bja/63.6.736

[3] Home I, Sutherland TD, Oakehott JG, Russel RJ. Cloning and expression of the phosphotriesterase gene hocA from pseudomonas monteilii C11. Microbiology (Reading, England). 2002;148(Pt 9):2687-2695. DOI: HYPERLINK "https://doi.org/10.1099/00221287-148-9-2687"10.1099/00221287-148-9-2687

[4] Matsuo M, Sekizawa J, Eto M. Fenitrothion. IPCS—International Programme on Chemical Safety. Environmental Health Criteria 133. Geneva: World Health Organization; 1992

[5] Kumar S, Kaushik G, Dar MA, Villarreal-Chiu J. Microbial degradation of organophosphate pesticides: A review. Journal in Pedosphere. 2018;28(2):190-208. DOI: HYPERLINK "http://dx.doi.org/10.1016/S1002-0160(18)60017-7"10.1016/S1002-0160(18)60017-7

[6] Pakala SB, Gorla P, Pinjari AB, Krovidi RK, Baru R, Yanamanandra M, et al. Biodegradation of methyl parathion and p-nitrophenol 2-hydroxylase in a Gram-negative Serratia sp. Strain DS001. Applied Microbiology and Biotechnology. 2007;**73**(6):1452-1462. DOI: 10.1007/s00253-006-0595-z

[7] Acharya KP, Shilpkar P, Shah MC, Chellapandi P. Biodegradation of insecticide monocrotophos by Bacillus subtilis KPA-1, isolated from agriculture soils. Applied Biochemistry and Biotechnology. 2015;**175**(4):1789-1804. DOI: 10.1007/s12010-014-1401-5

[8] Akhter MA, Laz R. Isolation and molecular characterization of pesticide (fenitrothion) resistant bacteria from agricultural field. IOSR Journal of Pharmacy. 2013;**3**(5):31-38. (e)-ISSN: 2250-3013, (p)-ISSN: 2319-4219. Available from: www.iosrphr.org

[9] Kumar M, Philip L. Bioremediation of endosulfan contaminated soil and water optimization of operating conditions in laboratory scale reactors. Journal of Hazardous Materials. 2006;**136**(2):354-364. DOI: HYPERLINK "https://doi.org/10.1016/j.jhazmat.2005.12.023"10.1016/j.jhazmat.2005.12.023

[10] Anwar S, Liaquat F, Khan QM, Khalid ZM, Iqbal S. Biodegradation of chlorpyrifos and its hydrolysis product 3,5,6-trichloro2-pyridinol by Bacillus pumilus strain C2A1. Journal of Hazardous Materials. 2009;**168**(1):400-405. DOI: 10.1016/j.jhazmat.2009.02.059

[11] Jayaraman P, Naveen Kumar T, Maheswaran P, Sagadevan E, Arumugam, P. In vitro studies on biodegradation of chlorpyrifos by Trichoderma viride and T. harzianum. Journal of Pure and Applied Microbiology. 2012;**6**(3):1465-1474. Available from: https://microbiologyjournal.org/in-vitro-studies-on-biodegradation-of-chlorpyrifos-by-trichoderma-viride-and-t-harzianum/

[12] Elias F, Woyessa D, Muleta D. Phosphate solubilization potential of rhizosphere fungi isolated from plants in Jimma Zone, Southwest Ethiopia. International Journal of Microbiology. 2016;**2016**:11. DOI: https://doi.org/10.1155/2016/5472601

[13] Omar SA. Availability of phosphorus and sulphur of insecticide origin by fungi. Biodegradation. 1998;**9**(5):327-336. DOI: 10.1023/a:1008310909262

[14] Kurtzman CP, Robnett CJ. Identification of clinically important ascomycetous yeasts based on nucleotide divergence in the 5' end of the large-subunit (26S) ribosomal DNA gene. Journal of Clinical Microbiology. 1997;**35**(5):1216-1223.

[15] Mukherjee I, Gopal M. Chromatographic techniques in the analysis of organochlorine pesticide residues, Journal of Chromatography A. 1996;**754**(1-2): 33-42. DOI: 10.1016/S0021-9673(96)00426-8

[16] Neti N, Zakkula V. Analysis of chlorpyrifos degradation by Kocuria sp. using GC and FTIR. Current Biotica. 2013;**6**(4):466-472. ISSN 0973-4031

[17] Liu S, Yao K, Jia D, Zhao N, Lai W, Yuan H. A pretreatment method for HPLC analysis of cypermethrin in microbial degradation systems. Journal of Chromatographic Science. 2012;**50**(6):469-476. DOI: 10.1093/chromsci/bms030

Chapter 9

# Insect Conservation and Management: A Need of the Hour

*Muzafar Riyaz, Rauf Ahmad Shah*
*and Soosaimanickam Maria Packiam*

## Abstract

Insects play a very vital role in divergent ecosystems and have gained great economic and medical importance as pollinators, pests, predators, parasitoids, decomposers and vectors. With the large-scale practice of synthetic pesticides, the diminishing rate of beneficial and pollinator insects is increasing rapidly. Environmental pollution, climate change, global warming, urbanization, industrialization and some natural calamities like wildfires add more fuel to the acceleration of insect decline all over the world. Alternative steps should be employed to replace the toxic pesticides and implementation of integrated pest management (IPM) should be put forward to reduce the overuse of synthetic pesticides and fertilizers, which have a great impact on beneficial insects as well as birds, aquatic organisms, and also on human health. The present study aims to create awareness among the researchers and general public by providing a brief review of insect importance, decline and conservation strategies.

**Keywords:** Insects, Pollinators, Insecticides, Climate Change, Insect decline, Conservation

## 1. Introduction

The most prevailing species ever to possess earth are Insects [1]. The amplified depiction of their body is a positive component to withstand in any environmental conditions. These six-legged creatures came to occupy the earth in the Devonian period and turned into the predominant animal's earth ever witness [2]. Unexpectedly, the insects ought to be appraised as exceedingly abundant creation, in light of the fact that with such an outfitted depiction of the body makes them dominating and the level of triumph achieved by a class of life frame inside invertebrate phyla [3]. With such a significant number of roles and the most noteworthy number of species in any population influences them prevailing life to shape on the earth. Insects are vital due to their diverseness, ecological character, and impact on farming, human wellbeing, and natural resources. Insects are viewed as the dominant animals on earth with their main competitors as humans. Humans have been relying upon the insects for the pollination of crops, honey, silk, lac and many other ecological services that insects provide in different ecosystems [4]. In an ecosystem, there are countless species of insects with their distinguished roles either associated with crops or other organisms in a particular location. The relationship of an insect with a crop or any other organism does not really imply that the species is a pest of that crop or animal. Most of the crops which needed pollination for their development are being pollinated by most of the insects, which are the prime agents of pollination among flowering plants [5]. Insects are very crucial for the appropriate functioning of many food chains and food webs. From nymphs of dragonflies as top predators of insect food-chains in aquatic ecosystems to grasshoppers, flies, butterflies and so on as primary consumers in many grassland biomes [6, 7]. Insects act as predators, parasitoids, herbivores, decomposers, sanguivores, parasites and also help in nutrient cycling. Insects play a very important role in decomposition which includes breakdown of waste, dead plant and animal matter, thus helps in remediation and recycling of our ecosystems [8]. The biological foundation for all terrestrial ecosystems is the insects with innumerable roles not limited to terrestrial ecosystems nevertheless they provide many useful services in and around the aquatic and agricultural ecosystems as well. Forensic and medical entomology involves the study and investigation of many insect species. From maggots of blowflies to larvae of mosquitoes, the advancement in the science of forensics and vector biology is only possible because of the deep investigation of these insect species which have changed the history of human intellectual. From the Devonian period to the present era of technological advancements, earth has witnessed these six-legged flying animals which dominated both the skies as well as the terrains [9]. As the over-use of synthetic pesticides, expansion of agriculture, urbanization, industrialization, environmental pollution, rising temperatures, climate changes came into existence, the insect species are becoming no longer the dominant animals on the planet and the risk of being threatened and receiving extinction is on the verge till this day [10].

The unending requirement of food for the fast-growing human population of the world has created havoc among the diversity of insects and other animals from different taxa by the manufacturing of toxic agrochemicals including pesticides sprayed on the crops for the eradication of pests [11]. The repeated use of these toxic pesticides sprayed in crop fields not only eradicate the pests, but also directly responsible for the decline of beneficial insects, which are having a great value to carry out the process of pollination and being as predators and parasitoids to check the diversity of insect pests in the natural ecosystem. Besides the damage done by the continuous application of synthetic pesticides on the insect biodiversity, there are many factors which are equally responsible for the insect decline. The fast-growing human population gave rise to the wide-spread expansion of urbanization, industrialization and assemblage of building and road network constructions which sequentially steered to the deforestation, habitat fragmentation and biodiversity loss. On contrary, climate change, rising temperatures, environmental pollutions are some of the main drivers of the global insect decline [12]. The introduction of alien and invasive plant species has also affected the insect diversity to some extant as the insects are mostly adapted to native plant and tree species. Implementation of conservation and management strategies of insects are need of the hour as the insect populations are falling at very higher proportions. The endangered and critically endangered insect species should be given top priority in terms of conservation. Additional insect surveys and field visits must be supported so that monitoring should be directed for proper analyzing and scrutinizing of endangered insect species. Comprehensive research studies, Citizen science projects could be implemented at a very large-scale, so that populations of insects and their diversity, richness and abundance can be monitored easily.

## 2. Importance of insects: a general concept

Insects are one of the dynamic groups of organisms in the kingdom animalia. The distinguished roles played by Insects in all biological systems makes them one of the prevailing class, earth at any point saw (**Figure 1**). The potential to withstand in any climatic condition, light weight, small size, flight capability makes them significantly versatile to endure and reproduce more faster than some other living forms on the planet. Insects were the first animals to ever develop the ability to fly. Since evolution usually works with what it has; new body structures do not crop up very often. However, in case of insects, they did not use modified limbs to fly. The

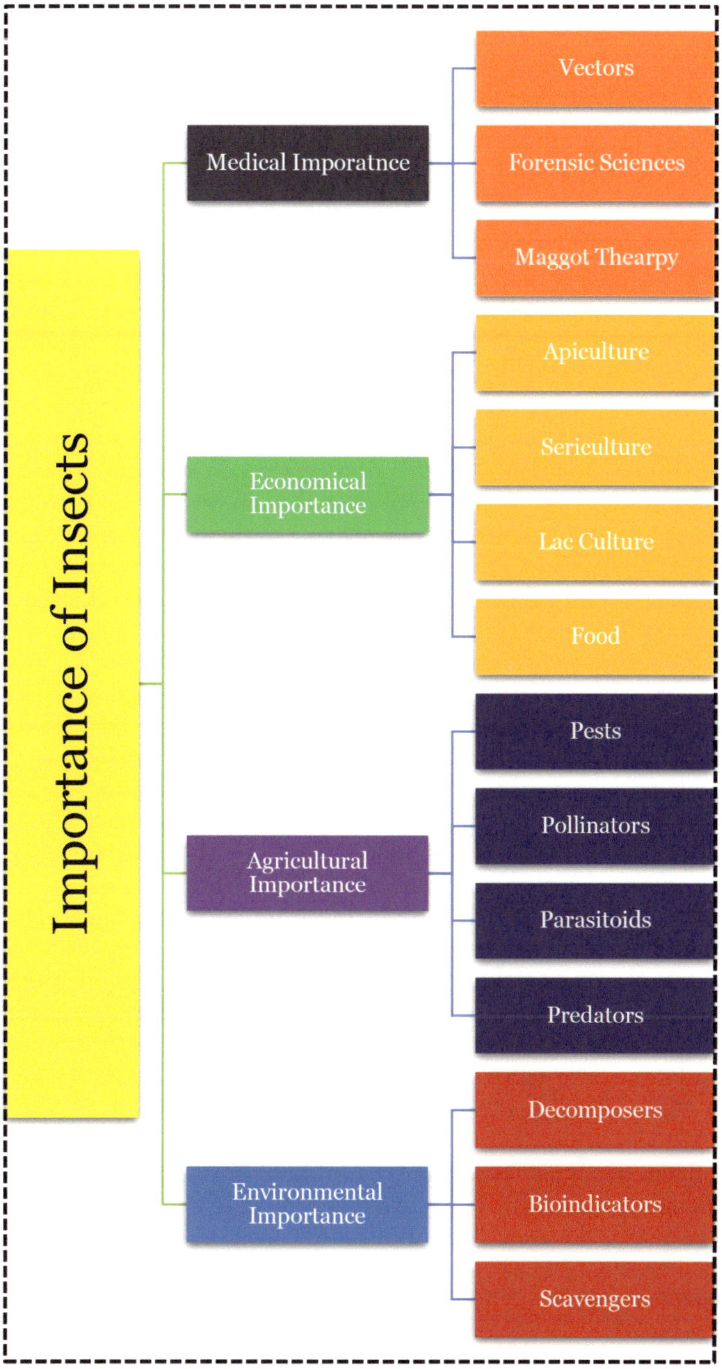


**Figure 1.**
*A flowchart showing the importance of insects (Designed in MS PowerPoint by Muzafar Riyaz)*

insect wings are a brand-new innovation in their physiology. The development of wings among them is so unusual that scientists are still working on, and arguing about how and when insect wings first came about. Nearly more than 1 million insect species have been discovered so far and scientists estimate that there could be million more waiting to be discovered. The faster reproductive rate, flight ability, light weight, unique body structures and major roles in different ecosystems makes them most dominant animals the earth has ever witnessed.

Insects play major roles in our environment however; insects are some of the most misunderstood and underappreciated animals on earth due to their capacity to destroy crops and carry diseases. Yet, insects are very crucial for better functioning of many ecosystems. One of the most important services that insects deliver is the pollination. Insects help in pollination of around 80% of the angiosperms across the globe [13]. Insects are very important in systematic functioning of many food chains and food webs as they provide food for many animals including birds, amphibians and reptiles. There are many significant assets that insects have been provided to Humans like Honey, Silk, Lac, Wax etc. Besides feeding on our crops and vegetables as pests, numerous insect species play crucial roles in eliminating many pest species as predators and parasitoids. Many predatory and parasitoid species of insects feed on Mosquitoes, aphids, pest caterpillars and mealybugs that destroy fruits and vegetable crops, therefore act as biological control agents in our ecosystems. Insects have been used in molecular and genetic studies, forensic sciences and many other biological studies including therapies. Many insects such as dragonflies act as biological indicators in the environment [14]. These species help in monitoring the biological quality of water as there are very sensitive to pollution. Most of the insect species help in environmental remediation as they spend most of their lives under water or inside soils. Insects play a very crucial role in the decomposition of plant and animal matter. The role of insects is so crucial that if insects and other land-dwelling arthropods were to become extinct, then it would sound death knell for all the earthlings. Majority of the birds, reptiles and mammals and amphibians would soon fizzle out to extinction. Next in line be the flowering plants, the physical structure of the forests and soon other terrestrial habitats will suffer an equatorial damage due to the disturbance in the food chains and food webs. Apart from ecosystem services, insects have been mentioned in folklores of many tribes and communities of peoples from all over the world. Many traditions across globe have considered insects as the treasures of the world. The ecosystem services delivered by insects on the planet are innumerable. However, due to some anthropogenic activities the populations of many insect species are rapidly running towards the engenderment. The large-scale utilization of synthetic pesticides has created a havoc among beneficial insect populations. Apart from synthetic pesticides, climate change, cryptic and alien plant and animal species are also responsible for the decline of insects.

## 3. Impact of anthropogenic activities on Insect diversity

One of the most common misconceptions about insects is the pest nature. Since, many of the species among different insect orders and families are pests however, not all of the insects are pests. A lot of this is based on the personal opinions of common people which need to be changed fundamentally by taking initiatives such as public awareness and citizen science. Global decline of insects is a very big problem that we are witnessing in the present era and a lot of people are not aware of what's happening and it's difficult to understand because the insects are seemingly everywhere. A lot of studies have revealed that the insects are disappearing at a high rate with estimates suggesting to 40% of the species in class Insecta will disappear in the couple decades [15]. The trend is pretty clear that insects are disappearing both in species and in number as well. The decline of insect biodiversity across the globe falls on many anthropogenic activities like habitat destruction through deforestation, hunting, expansion of agriculture, industrialization and urbanization. Large-scale intensification of agricultural activities has resulted in decline of populations among the insects. The enormous utilization of synthetic pesticides is a result of the expansion of agricultural activities and adds as one of the top drivers of insect population decline (**Figure 2**). Besides the impact of synthetic pesticides on insects, many other factors are also responsible for their decline. Destruction of pond and wetland habitats, increasing temperatures, introduced species, ecological traits, pollution, wildfires are some of the key factors which are associated with the decline of insect populations across the globe (**Figure 3**). According to many reports, order Coleoptera (Beetles and weevils) is being highly affected by the habitat change followed by orders Hymenoptera (Bees and Wasps), Lepidoptera (Butterflies and Moths), Odonates (Dragonflies and Damselflies) and other terrestrial and aquatic insects. Pollution, climate change and biological traits are one of the main drivers associated with the vastly declining of insect species from the order Coleoptera followed by Hymenoptera, Lepidoptera, Odonata and other group of insects. These factors have caused a very huge damage to insect populations. As revealed by many studies across the globe, a very large of insect species such as Dung beetles followed by the bees, moths and butterflies are vulnerable and rapidly heading towards the endangerment. A decline of over <30% proportions of particular insect order can be seen among the Coccinellid beetles followed by the orthopterans, butterflies, hymenopterans and in case of aquatic insect species from the order Odonata followed by Ephemeroptera, Plecoptera and Trichoptera [16–19].

Global warming and climate change are equally responsible for the decline of insect populations. The insects of temperate regions across the globe are among the most affected species of insects. Insect species such as dragonflies, stoneflies and bumblebees which are adapted to cold climates and higher altitudes are being affected by the rising temperatures in temperate regions of the world. Besides, the

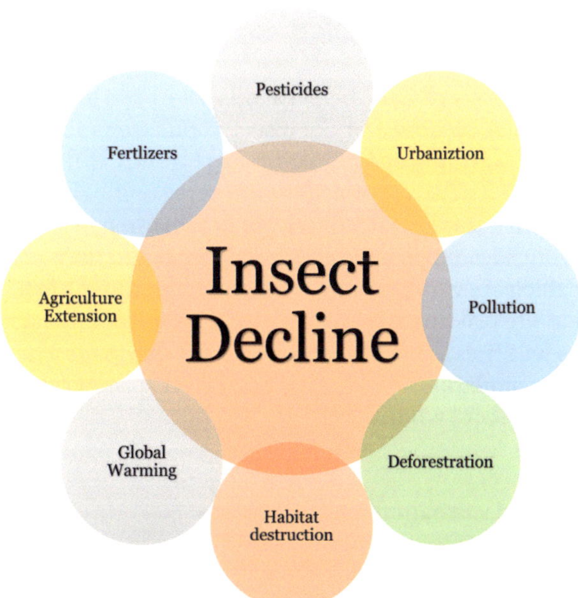


**Figure 2.**
*Drivers of insect decline (Designed in MS PowerPoint by Muzafar Riyaz).*

**Figure 3.**
*Impact of synthetic agrochemicals on different species of insects. (Photos by Muzafar Riyaz).*

insects from the rainforests of Caribbean islands have been drastically affected by the climate change. Almost half of the insect populations across the globe are affected by the global warming and climate change trends [20]. Other factors that are equally responsible for the insect decline are persistent halogenated hydrocarbons, metal pollution, heavy metals. These pollutants often discharged into rivers, lakes and ponds which lead to in an innumerable impact to the aquatic insect fauna. Industrial spills which are very toxic not only affect the aquatic insect fauna but also other forms of life residing in both fresh and salt waters. On contrary, natural calamities such as wildfires, cyclones and so on have also made a huge impact on the reduction of insect populations. Many endemic insect species are believed to face extinction due to the recent wildfires in Australia.

## 4. Conservation and management of insects

Insect decline is very complicated as it is been driven by many anthropogenic and natural activities. Th populations of insects are disappearing at an alarming rate and the total mass of insects is falling by a staggering 2.5% a year [21]. Insect species such as beetles, ants, bees are disappearing eight times faster than mammals, birds or reptiles. The population of monarch butterflies in the United States reduced by 90% in the last 20 years [22]. Insects outweigh every other animal and make up around 70% of all animal species on the planet, however there are reports of widespread decline of insect species from every corner of the world. Since, the humans have been worried about the bees for a while however, the concern for overall insect decline is much bigger than just the bees.

Insect decline is indeed problematic that we need to tackle as these animal species are very important for proper functioning of our ecosystems. The implementation of conservation and management strategies is a need of the hour. The following steps need to be implemented for conservation of the insect species (**Figure 4**):

a. Native plant species should be given the importance as compared to invasive plant species, since most of the insects rely on the friendly plantations around them.

b. The rapidly growing urbanization and industrialization must be designed in eco-friendly ways.

c. Alien and cryptic plant and tree species should be removed as most of the insect species are poorly adapted to these plant species.

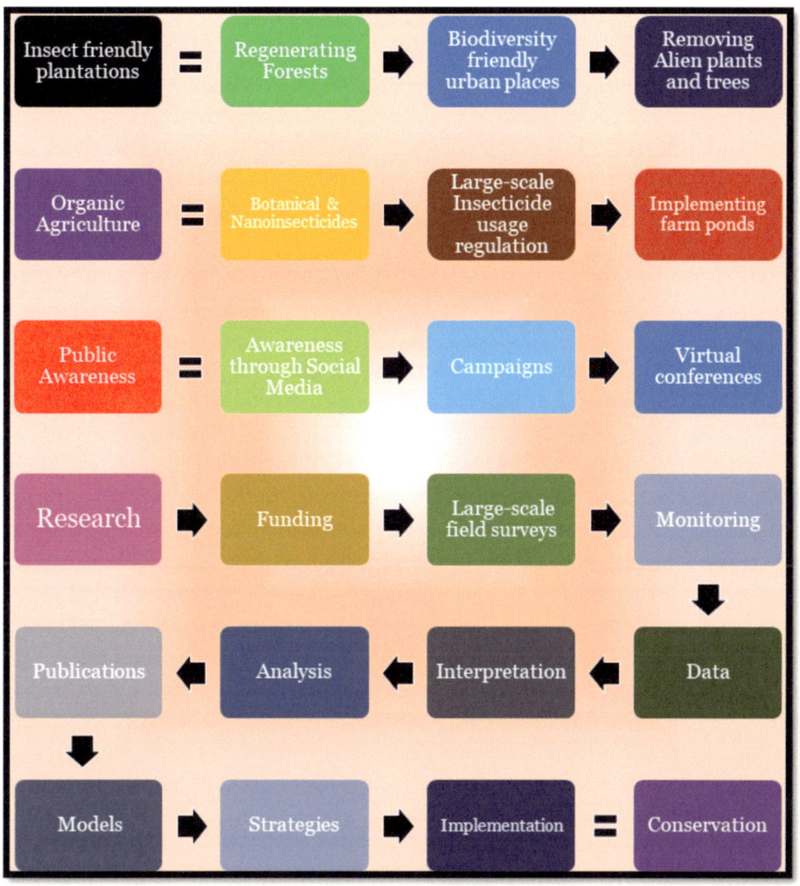


**Figure 4.**
*An overview of conservational and management strategies of insects. (Designed in MS PowerPoint by Muzafar Riyaz).*

d. Deforestation is one of the major issues in terms of the biodiversity loss and insect decline. Regeneration of the forests is a need of the hour, since forests are the common home to all wild fauna including the insects.

e. The large-scale extension of agricultural activities must be regulated and managed in a proper channel. Organic farming should be implemented in way which support crop yield as well the biodiversity around.

f. With the use of botanical pesticides, the extensive utilization of synthetic pesticides can be controlled to some extent.

g. With nanotechnological approaches like use of nanopesticides and nanofertilizers will help in reducing the wide-spread utilization of insecticide pollution.

h. Use of synthetic pesticides which not only affects the life from different taxa but costs the human health as well [23]. Alternative to these synthetic pesticides are the biological control agents such as insect predators and parasitoids which are also the core component of ecological intensification in Integrated Pest Management (IPM) [24, 25].

i. Implementing small farm ponds in the agricultural fields will aid in restoration of aquatic insect taxa like dragonflies and stoneflies.

j. Public awareness is one of the key factors in conserving and managing the insect decline. In these times of technology and fast-growing civilizations, people are ignoring the ecological services of nature and natural ecosystems. People should be made aware of the importance of insect diversity and our biodiversity wealth and their conservation and management.

k. Public awareness should be made through campaigns, seminars, conferences about the insect diversity and conservation.

l. Rather than sharing memes on the social media, general people must be trained for sharing the benefits of insects and their conservation aspects.

m. Citizen science aims at increasing scientific knowledge through collaboration and public participation in scientific research. With this initiative, people from different parts of the world take part in such activities and share and contribute to data monitoring and collection programmes. Citizen science allows people to enhance their scientific temperament in the fields and empowers communities to observe nature and with the collective efforts to conserve as well. Across the globe, majority of people are been taking part in such activities and people are very enthusiastic about sharing and collaborating to scientific research. More citizen science projects must be initiated in the future as well so that more discoveries of species can be made and together with the public, scientists and researchers will be able to solve the insect decline problems.

n. Diversity studies of the most of the insect orders have been ignored in case of Collembola, Ephemeroptera, Neuroptera, Plecoptera which are having many ecological roles and services.

o. Monitoring should be carried out for each and every insect species, so that data should be utilized in the conservational strategies.

p. Research is one of the main aspects for monitoring both the insect decline and cause of the decline. Researcher's and Scientists across the globe should be invigorated to study some of the major aspects of insect decline which have not been adequately studied. These include, impact of industrial chemicals on insects, heavy metals, thermal biology etc.

q. Funding is a very core aspect in surveying, monitoring, data analysis and laboratory studies. A lot of funding agencies across the globe play an important role in conservation and management of the animals. Focus should be also given to insects which are heading towards the apocalypse.

r. Journals and publishers must process faster in reviewing the articles, papers, chapters concerning the insect conservation and management and should be made freely accessible to all.

## 5. Discussion and conclusion

Insects perform all sorts of important ecological roles without which ecosystems could not function. The biodiversity crisis in the present era has resulted in the loss of species from our planet faster than has happened for 65 million years, since the dinosaurs were wiped out by a meteor. The perception about the conservation for most of the people is that it's about large animals like tigers, pandas, polar bears and so on and that's what where most of the attention goes and trying to prevent those creatures from going extinct. However, while focusing on the mammalian and other species conservation we have missed the bigger picture that is been going on in our environment which is the quite disappearance of the insects. The disappearance of the insect species has been going on for a long time. Insect biodiversity needs to be preserved in order to preserve both the flora and fauna of the earth. Biodiversity has been very important to human history and culture as humans are totally depend on both plants and animals which live around them for food as well as for cultural value and they make our ecosystems healthy in which humans take shelter and yet so much of it is under threat. Nature has a lot of value and biodiversity is the basis of life. One of the biggest consequences of a more developed and more technological world is that people flock to cities which resulted in making humans more gentrified and more separate from the nature. As we are aware of the fact that insect biodiversity is declining dramatically all across the globe. It is very important that we have information management systems to know what's happening and what drivers are causing the insect decline, so that management strategies should be implemented for the conservation of the entomofauna. The importance of the insect fauna cannot be over-emphasized as it is very important for proper balancing of our ecosystems and ecosystems services they provide. Insects are fueling a wide range of ecosystems services that we essentially need as humans to survive. However, it is very important that even before we can save them, we need to get to know about them. The better and advanced decisions are needed in these times of insect biodiversity loss and much care is needed of all the insect fauna that are in threat to become endangered or extinct. The knowledge about these species is very important for their conservation and management. The above-mentioned steps need to be implemented as far as we can, so that our future generations will get to see the natural heritage of our planet. Ultimately biodiversity will become important once it means something to each and every individual.

## Acknowledgements

The authors wish to thank Entomology Research Institute for extended guidance and support and aims to create a better world through conservation activities, research, publications and collaborations.

## Conflict of interest

The authors declare no conflict of interest.

## Author details

Muzafar Riyaz, Rauf Ahmad Shah and Soosaimanickam Maria Packiam*
Entomology Research Institute, Loyola College, Chennai, Tamil Nadu, India

*Address all correspondence to: eripub@loyolacollege.edu

## References

[1] Glenner H, Thomsen PF, Hebsgaard MB, Sørensen MV, Willerslev E. The origin of insects. Science. 2006 Dec 22;314(5807): 1883-1884.

[2] Carpenter FM. The geological history and evolution of insects. American Scientist. 1953 Apr 1;41(2):256-270.

[3] White TC. Insects. In: The Inadequate Environment, White, T.C. (Ed.). 1993 (pp. 31-107). Springer, Berlin, Heidelberg.

[4] Rosenberg DM, Danks HV, Lehmkuhl DM. Importance of insects in environmental impact assessment. Environmental management. 1986 Nov;10(6):773-783.

[5] Riyaz M, Mathew P, Paulraj G, Ignacimuthu S. Entomophily of Apple ecosystem in Kashmir valley, India: A review. International Journal Scientific Research in Biological Sciences. 2018; 5(5): 146-154.

[6] Pimm SL. Properties of food webs. Ecology. 1980 Apr;61(2):219-225.

[7] Pimm SL, Lawton JH. Are food webs divided into compartments? The Journal of Animal Ecology. 1980 Oct 1:879-898.

[8] Simmons T, Cross PA, Adlam RE, Moffatt C. The influence of insects on decomposition rate in buried and surface remains. Journal of Forensic Sciences. 2010 Jul;55(4):889-892. https://doi.org/10.1111/j.1556-4029.2010.01402.x

[9] Wigglesworth VB. Evolution of insect wings and flight. Nature. 1973 Nov;246(5429):127-129. https://doi.org/10.1038/246127a0

[10] Leather SR. "Ecological Armageddon"-more evidence for the drastic decline in insect numbers. Annals of Applied Biology. 2017 Dec 20;172(1):1-3. https://doi.org/10.1111/aab.12410

[11] Williams CM. Third-generation pesticides. Scientific American. 1967 Jul 1;217(1):13-17.

[12] Wilson RJ, Maclean IM. Recent evidence for the climate change threat to Lepidoptera and other insects. Journal of Insect Conservation. 2011 Apr;15(1):259-268. https://doi.org/10.1016/j.foreco.2014.05.027

[13] Faheem M, Aslam M, Razaq M. Pollination ecology with special reference to insects a review. J Res Sci. 2004;4(1):395-409.

[14] Riyaz M. Dragonflies: The Apex Predators of the Insect World. Academia Letters. 2021:1-4. Article 1365. https://doi.org/10.20935/AL1365.

[15] Wagner DL, Grames EM, Forister ML, Berenbaum MR, Stopak D. Insect decline in the Anthropocene: Death by a thousand cuts. Proceedings of the National Academy of Sciences. 2021 Jan 12;118(2). https://doi.org/10.1073/pnas.2023989118

[16] Sánchez-Bayo F, Wyckhuys KA. Worldwide decline of the entomofauna: A review of its drivers. Biological conservation. 2019 Apr 1;232: 8-27. https://doi.org/10.1016/j.biocon.2019.01.020

[17] van der Sluijs JP. Insect decline, an emerging global environmental risk. Current Opinion in Environmental Sustainability. 2020 Oct 24. https://doi.org/10.1016/j.cosust.2020.08.012

[18] Goulson D. The insect apocalypse, and why it matters. Current Biology. 2019 Oct 7;29(19): R967-R971. https://doi.org/10.1016/j.cub.2019.06.069

[19] Saunders ME. Ups and downs of insect populations. Nature ecology &

evolution. 2019 Dec;3(12):1616-1617. https://doi.org/10.1038/s41559-019-1038-4

[20] Halsch CA, Shapiro AM, Fordyce JA, Nice CC, Thorne JH, Waetjen DP, Forister ML. Insects and recent climate change. Proceedings of the national academy of sciences. 2021 Jan 12;118(2). https://doi.org/10.1073/pnas.2002543117

[21] Montgomery GA, Dunn RR, Fox R, Jongejans E, Leather SR, Saunders ME, Shortall CR, Tingley MW, Wagner DL. Is the insect apocalypse upon us? How to find out. Biological Conservation. 2020 Jan 1;241:108327. https://doi.org/10.1016/j.biocon.2019.108327

[22] Wepprich T, Adrion JR, Ries L, Wiedmann J, Haddad NM. Butterfly abundance declines over 20 years of systematic monitoring in Ohio, USA. PLoS One. 2019 Jul 9;14(7):e0216270. https://doi.org/10.1371/journal.pone.0216270

[23] Riyaz M, Shah RA, Sivasankaran K. Pesticide Residues: Impacts on Fauna and the Environment. In: Biodegradation, Mendes, K. F. (Ed.). 2021. (pp. 1-13). IntechOpen. http://dx.doi.org/10.5772/intechopen.98379

[24] El-Shafie HA. Integrated Insect Pest Management. In: Pests Control and Acarology, Haouas, D and Hufnagel, L. (Eds.). 2018 Dec 31. (pp. 1-15). IntechOpen. https://doi.org/10.5772/intechopen.81827

[25] El-Shafie HA. Insect pest management in organic farming system. In: Multifunctionality and Impacts of Organic and Conventional Agriculture, Moudrý, J., Bernas, J., and Mendes, K. F. (Eds.). 2019 Mar 15. (pp. 1-13). IntechOpen. https://doi.org/10.5772/intechopen.84483

Chapter 10

# Description of a New Species of the Genus *Anagrus* (Hymenoptera: Chalcidoidea: Mymaridae): A Biocontrol Agent as an Alternative to Insecticide Use

*Shireen Saleem and Shoeba Binte Anis*

## Abstract

Although insects are economically important as they produce honey, silk, act as pollinators and also play an important role in functioning of an ecosystem, yet insect population is declining very fast. One of the possible causes of insects decline is excessive use of pesticides. Control of pest with synthetic chemicals or pesticides result in several issues and complications. These chemical pesticides or insecticides can also cause toxic effects on beneficial organisms like honeybees and butterflies which are important pollinators. So, biocontrol agents can be used as best alternative to control pest without harming beneficial organism and non-target insects or other organism as majority of biocontrol agents are host specific. Biological control agents including predators and parasotoids are natural enemies of insect pests. Present chapter deals with the description and illustration of one new species *Anagrus (Anagrus) sololinearis* sp.nov from India. This new species belongs to genus *Anagrus* (Hymenoptera: Chalcidoidea: Mymaridae). Genus *Anagrus* is considered as one of the important and most promising biocontrol agents in insects as it is an egg parasitoid.

**Keywords:** *Anagrus*, biocontrol agent, new species

## 1. Introduction

Insects belonging to phylum Arthropoda are the most biodiverse group of fascinating creatures and can be found in aquatic as well as terrestrial habitats. Although insects are economically important and are key pollinators, yet they are declining at global level. Several studies have been carried out in different regions which reported a substantial decline in insect populations. Several researchers studied insects decline and possible causes of their decline at global level. Recent studies, reviews and causes of insects decline were mainly based on researches from the United States or Europe. A group of European researchers in October 2017 reported that insect abundance had declined by more than 75% within 63 protected areas in Germany over the course of 27 years [1]. Stork [2]; Habel et al. [3]; Forister et al. [4]; Bayo & Wyckhuys [5]; Wagner [6]; Eggleton [7]; Klink et al. [8] and Wagner et al. [9] made noteworthy and remarkable contributions regarding the review and study of insects decline and



IntechOpen

causes of decline at global level. Insect's populations are being declining at various rates across space and time, the decline in abundance on an average is thought to around 1–2% per year. Loss of insect diversity and abundance is expected to provoke cascading effects on food webs and to jeopardize ecosystem services [1].

Insects play a very important role in food chain and food web of an ecosystem. Butterflies and bees are considered as good pollinators. Termites and dung beetles act as decomposers. Insect's products like honey and silk are commercially important. There is an unending list of insect's economic importance and key role in ecosystem and therefore, their decline is a matter of concern and there is also a great need to find the causes of decline. There are various causes of insects decline. Some possible causes of insects decline include intensive farming, urbanization, change in climate as well as use of pesticides. Excessive use of pesticides including insecticides on agricultural crops can be toxic to a host of other organisms including beneficial insects as well as other non-target species. Pesticides have severe impact on environment too [10]. Integrated pest Management (IPM) combines the use of biological, cultural and chemical practices in agriculture to control pests. It focuses on use of natural predators, parasites and parasitoids. IPM is the best approach as it sustainably manages insects by focusing mainly on prevention rather than treatments and without doubt, it is also an environment friendly approach.

Biological pest control, an important method of IPM involves the use of another living organism to kill a pest. As no chemicals are involved, therefore no environment contamination occurs as it happens with use of chemical pesticides. One of the advantages of biological pest control also lies in the fact that the pests do not develop resistance against biocontrol agents. Biological control agents including predators and parasotoids are natural enemies of insect pests. Order hymenoptera of class Insecta form an extremely diverse group with over 1, 15,000 described species comprising almost 10% of the species diversity on the earth [11]. The order Hymenoptera includes sawflies, bees, ants and wasps, and together they directly affect human health and agriculture through diverse roles such as pollinators, pests and parasitoids [12]. The Chalcidoidea is a large hymenopteran superfamily, the majority of which are entomophagous parasitoids with hosts in a wide range of insect orders [13, 14]. Family Mymaridae belonging to superfamily Chalcidoidea includes the smallest known insects, all parasitoids in the eggs of other insects [15] except for two that parasitize larvae of a species of family Eulophidae [16]. So far, many insect species have been successfully used as biocontrol agents against various pests on agriculturally important crops. Biocontrol agents can be used as best alternative to control pest without harming beneficial organism and non-target insects or other organism as majority of biocontrol agents are host specific.

One of the important and most promising biocontrol agents in insects is genus *Anagrus* which is an egg parasitoid. Many of its species have been used successfully to control leafhoppers on apple, rice & grape [17–19]. Prior to use as biocontrol agent in integrated pest management, correct identification at generic as well as at species level is a very necessary step. Taxonomy basically deals with the identification and classification. Present work includes the description and illustration of a new species *Anagrus (Anagrus) sololinearis* sp.nov. of promising biocontrol agent genus *Anagrus* (Hymenoptera: Chalcidoidea: Mymaridae) from India.

## 2. Material and methods

The insect specimens collected by sweeping, mounted on cards, and after card mounting, slides were prepared by adopting the procedure given by Noyes [20].

Body color was noted down from the card-mounted specimen. Only body length was taken from card mounted specimen and is given in millimeters (mm). Other measurements (of slide mounted specimens) are relative, and were taken from the divisions of a linear scale of a micrometer placed in the eyepiece of a compound microscope Nikon Eclipse E200. These measurements were taken at 400 × magnification (1 division = 0.00274 mm) of the microscope.

Photographs of slide mounted specimens were taken by the digital camera "Leica, DFC295" fitted over a compound microscope (Leica, DM2500). Line diagrams were made using Nikon Eclipse 80i at 400 × at zoom 9 and 11.

The following abbreviations were used:

F1, F2 and F3 = funicle segments 1, 2, 3 etc. of antenna.
OOL = minimum distance between a posterior ocellus and an eye margin.
POL = minimum distance between the two posterior ocelli.
FWL = Fore wing length.
FWW = Fore wing width.

The following acronym is used for the depository:

ZDAMU = Insect Collections, Department of Zoology, Aligarh Muslim University, Aligarh, India.

## 3. Results and discussion

### 3.1 *Anagrus* Haliday

*Anagrus* Haliday, 1833: 346. Type species *Ichneumon atomus* Linnaeus, 1767:941, designated by Westwood, 1840:78 [21, 22].

Brief diagnosis: Female antennal clava entire, scape with transverse folds; each mandible tridentate. Axillae of mesosoma advanced into side lobes of mesoscutum. Forewing with posterior margin (behind venation) only slightly lobed. Posterior scutellum short and divided by a longitudinal sulcus in two lobes. Posterior scutellum about as long as or slightly longer than anterior scutellum. Foretibial spur comb-like [23–26].

### 3.2 *Anagrus (Anagrus) sololinearis* sp.nov.

#### *3.2.1 Description*

Length (excluding exserted ovipositor). 0.40 mm. Body light yellow. Head yellowish; eyes black. Antenna pale brown. Fore and hind wing hyaline. Legs light yellow. Gaster yellowish brown, posterior two-third part of gaster blackish brown (**Figure 1**(**1–4**)).

##### *3.2.1.1 Head*

Almost triangular in frontal view, 1.7 × as broad as high (82:46); OOL 1.5 × POL (12:8); eye height about 2 × as long as malar space (37:18). Mandible brown, tridentate (**Figure 1**(**1**)). Antenna (**Figure 1**(**2**)) with scape swollen ventrally, 3.5 × as long as broad; pedicel 2 × as long as broad, 2.6 × as long as F1; F1 small, globular; F2 slightly shorter than following funicular segments; F3 and F5 exactly equal in length; F4 and F6 equal in length; F3- F4 each with 1 longitudinal sensillum; F5 without longitudinal

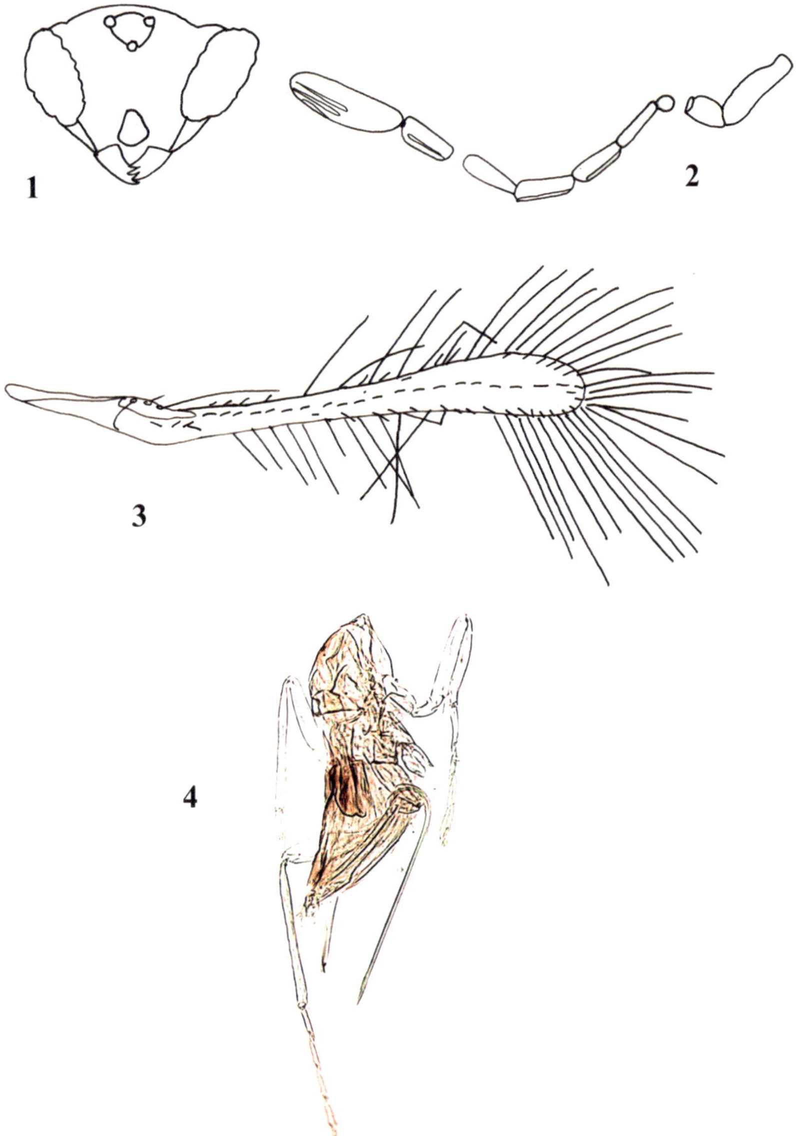


**Figure 1.**
*(1–4) Anagrus (Anagrus) sololinearis sp.nov. Female: 1, head; 2, antenna; 3, fore wing; 4, body (mesosoma and metasoma).*

sensillum; F6 with 1 longitudinal sensillum; clava 3.5 × as long as broad, slightly longer than combined lengths of F5 and F6; clava with 3 longitudinal sensilla (**Figure 1**(**1**)).

### *3.2.1.2 Mesosoma*

Mid lobe of mesoscutum without adnotaular setae. Mesoscutum with distinct notauli. Fore wing (**Figure 1**(**3**)) 9.2 × times as long as broad; forewing disc with bare area and with only 1 median row of setae in broadest part; marginal fringe about 3 × the wing width; distal and proximal macrochaetae in ratio 5.2:1 (**Figure 1**(**4**)).

#### *3.2.1.3 Metasoma*

Slightly longer than mesosoma, about 1.1 × as long as mesosoma length (80: 70); ovipositor strongly overlapping mesophragma anteriorly and posteriorly slightly exserted beyond apex of gaster; ratio of total ovipositor length to length of its exserted part 4.0:1; external plates of ovipositor each bearing 1 seta; ovipositor 2.3 × as long as fore tibia length (**Figure 1(4)**).

#### *3.2.1.4 Relative measurements (on slide)*

Scape length, 32; scape width, 9; pedicel length, 16; pedicel width, 8; F1, 6; F2, 17; F3, 20; F4, 21; F5, 20; F6, 21; clava length, 43; clava width, 12; FWL, 212; FWW, 23; marginal fringe, 70; distal macrochaeta length, 37; proximal macrochaeta length, 7; fore tibia length, 48; ovipositor length, 114; exserted ovipositor length, 28.

#### *3.2.1.5 Material examined*

Holotype, female (on slide). INDIA: ORISSA [=ODISHA]: Puri Matia Pada, 1.xii.2007, coll. FR Khan (ZDAMU).

Paratypes, 4 females: 1 female (on slide, same data as holotype) (ZDAMU). 3 females (on slides). INDIA: ODISHA = ORISSA: Pur Chandanpur, 29.xi.2007, coll. FR Khan (ZDAMU).

#### *3.2.1.6 Etymology*

The species name based on single row or line of setae present on fore wing.

#### *3.2.1.7 Hosts*

Unknown.

#### *3.2.1.8 Distribution*

India: Odisha.

#### *3.2.1.9 Male*

Unknown.

## 4. Comments

This new species belongs to "*atomus*" species group of *Anagrus* s. str., and can be distinguished from other species of *atomus* group by its unique combination of characters i.e. presence of longitudinal sensilla on F3 & F4; F5 without longitudinal sensillum; bare area present on fore wing disc; fore wing disc with only one median row of setae. *A. (A.) sololinearis* sp. nov. is similar to *A. (A.) frequens* Perkins in having fore wing disc with bare area and F4 with 1 longitudinal sensillum but differs from it in the following characters: F5 without longitudinal sensillum; only one median row of setae present on fore wing disc; fore wing about 9.2 × as long as broad; ratio of total ovipositor length to length of its exserted part 4.0:1. In *A (A.) frequens*, F5 with longitudinal sensillum; 2 rows of setae present on forewing disc;

fore wing more than 10.5× as long as broad; ratio of total ovipositor length to length of its exserted part more than 5.0:1.

## 5. Discussion

In the present work, a new species *Anagrus (Anagrus) sololinearis* sp.nov. belonging to genus *Anagrus* (Hymenoptera: Chalcidoidea: Mymaridae) was described and illustrated from India. Genus *Anagrus* is considered as most promising biocontrol agent against various insect pests as many of its species have been used successfully to control leafhoppers on apple, rice & grapes [17–19]. Minute fairy fly insect *Anagrus* can serve as best alternative to insecticide use if there is a correct identification of species of this parasitoid as well as its host.

## 6. Conclusion

Present work gives a brief idea about the role of insects as important components of an ecosystem as well as beneficial on a commercial basis by producing honey and silk. Due to such great importance of insects, their decline at global level is a cause of concern. Several studies by researchers carried out at global level confirmed the decline of these important fascinating creatures in different regions at varying rates to some extent. There is a need to find out the possible causes of insects decline. Excessive use of pesticides including insecticides on agricultural crops is also a cause and can be toxic to a host of other organisms including beneficial insects as well as non-target species. Pesticides can also have severe impact on environment. The present study also emphasizes on preference of biocontrol agents over pesticides or insecticides use. Biocontrol agents can be used as best alternative to control pest without harming beneficial organism and non-target insects or other organism as majority of biocontrol agents are host specific.

One of the important and most promising biocontrol agents in insects is genus *Anagrus* (Hymenoptera: Chalcidoidea: Mymaridae) which is an egg parasitoid. In the present work, a new species *Anagrus (Anagrus) sololinearis* sp.nov. from India is identified, described and illustrated. This species belongs to genus *Anagrus* (Hymenoptera: Chalcidoidea: Mymaridae). Genus *Anagrus* is an important egg parasitoid and promising biocontrol agent.

## Acknowledgements

Authors are thankful to Dr. Mohammad Hayat, Department of Zoology, Aligarh Muslim University, Aligarh for providing research material. The authors are also thankful to the Chairman, Department of Zoology, Aligarh Muslim University, Aligarh for providing working facilities.

## Author details

Shireen Saleem* and Shoeba Binte Anis
Section of Entomology, Department of Zoology, Aligarh Muslim University, Aligarh, Uttar Pradesh, India

*Address all correspondence to: shireen.ento@gmail.com

## References

[1] Hallmann, CA, et al. More than 75 percent decline over 27 years in total flying insect biomass in protected areas. PLoS ONE. 2017; 12(10): 1-21.

[2] Stork, NE. How many species of insects and other terrestrial arthropods are there on Earth? Annu. Rev. Entomol. 2018; 63: 31-45.

[3] Habel, JC, et al. Agricultural intensification drives butterfly decline. Insect Conserv. Divers. 2019; 12: 289-295.

[4] Forister, ML; Pelton, EM; Black, SH. Declines in insect abundance and diversity: We know enough to act now. Conserv. Sci. Pract. 2019; 1:e80.

[5] Sanchez-Bayo, F; Wyckhuys, KAG. Worldwide decline of the entomofauna: A review of its drivers. Biol. Conserv. 2019; 232:8-27.

[6] Wagner, DL. Insect declines in the Anthropocene. Annu. Rev. Entomol. 2020; 65: 457-480.

[7] Eggleton, P. The State of the World's Insects. Annual Review of Environment and Resources. 2020; 45(1): 8.1-8.22.

[8] Klink, R. van, et al. Meta-analysis reveals declines in terrestrial but increases in freshwater insect abundances. Science. 2020a; 368: 417-420.

[9] Wagner, DL, et al. Insect decline in the Anthropocene: Death by a thousand cuts. Proceedings of National Academy of Sciences of the United States of America. 2021; 118 (2):1-10.

[10] Gyawali, K. Pesticide Uses and its Effects on Public Health and Environment. Journal of Health Promotion. 2018; 6: 28-36.

[11] Martin, L. (2019, May 24). "Hymenopteran". Encyclopedia Britannica. Accessed 24 May, 2021 from https://www.britannica.com/animal/hymenopteran.

[12] Munoz-Torres, MC, et al. Hymenoptera Genome Database: integrated community resources for insect species of the order Hymenoptera. Nucleic Acids Research. 2011; 39 (Database issue): D658-D662.

[13] Heraty, JM, et al. A phylogenetic analysis of the megadiverse Chalcidoidea (Hymenoptera). Cladistics. 2013; 29(5): 466-542.

[14] Noyes, J.S. 2021. Universal Chalcidoidea Database. World Wide Web electronic publication. Accessed 20 March, 2021 from http://www.nhm.ac.uk/chalcidoids.

[15] Huber, JT. Systematics, biology, and hosts of the Mymaridae and Mymarommatidae (Insecta: Hymenoptera): 1758-1984. Entomography. 1986; 4: 185-243.

[16] Huber, JT, et al. Two new Australian species of *Stethynium* (Hymenoptera: Mymaridae), larval parasitoids of *Ophelimus maskelli* (Ashmead) (Hymenoptera: Eulophidae) on Eucalyptus. Journal of Natural History. 2006, 40: 1909-1921.

[17] Chiappini, E, et al. Key to the Holarctic species of *Anagrus* Haliday (Hymenoptera Mymaridae) with a review of the Nearctic and Palearctic (other than European) species and descriptions of new taxa. Journal of Natural History. 1996; 30(4): 551-595.

[18] Triapitsyn, SV & Teulon, DA.J. On the identity of *Anagrus* (Hymenoptera: Mymaridae) egg parasitoids of Froggatt's apple leafhopper, *Edwardsiana crataegi* (Douglas) (Homoptera: Cicadellidae), in

*DOI: http://dx.doi.org/10.5772/intechopen.99957*

Christchurch, New Zealand. New Zealand Entomologist. 2002; 25 (1): 91-92.

[19] Agboka, K, et al. (2004). Life-table study of *Anagrus atomus*, an egg parasitoid of the green leafhopper *Empoasca decipiens*, at four different temperatures. Bio Control. 2004; 49(3): 261-275.

[20] Noyes, JS. Collecting and preserving chalcid wasps (Hymenoptera: Chalcidoidea). Journal of Natural History. 1982; 16:315-334.

[21] Haliday, A. H. 1833. An essay on the classification of the parasitic Hymenoptera of Britain, which correspond with the Ichneumones minuti of Linnaeus. – Entomological Magazine 1: 259-276, 333-350.

[22] Westwood JO (1840) Synopsis of the genera of British insects: 1-154 Addenda to the generic synopsis of British insects.

[23] Schauff, ME.The holarctic genera of Mymaridae (Hymenoptera: Chalcidoidea). Memoirs of the Entomological Society of Washington. 1984; 12: 1-67.

[24] Yoshimoto, CM. A review of the genera of New World Mymaridae (Hymenoptera: Chalcidoidea). Flora & Fauna Handbook No. 7, Sandhill Crane Press, Inc., Gainesville, Florida.1990; 166.

[25] Huber, JT; Viggiani, G and Jesu, R. (2009). Order Hymenoptera, family Mymaridae. In: Arthropod fauna of the UAE. 2009; Volume 2. Harten, A. (Ed.): 290-297.

[26] Pricop, E. (2013). Identification key to European genera of the Mymaridae (Hymenoptera : chalcidoidea), with additional notes. ELBA Bioflux. 2013; 5: 69-81.

Chapter 11

# Impacts of Organic Farming on Insects Abundance and Diversity

*Hamadttu Abdel Farag El-Shafie*


## Abstract

Organic farming encourages maximum utilization of the natural biological processes to manage the farm in terms of soil fertilization and pest control, which implies using none or less synthetic fertilizers, pesticides, and plant and animal growth-promoting substances. All these practices increase arthropod diversity, particularly soil-dwelling insects. Intercropping, cover crops, and hedges, which are common practices in organic fields, provide alternative habitats for arthropod communities. The refugia also provide a good source of food for pollinators in terms of pollen grains and nectar. The interactions among the different plant and animal taxa (weeds, birds, mammals) that are found in the organic farming ecosystem have a great impact on insects' abundance and diversity. This chapter summarizes the impacts of the organic farming system on the abundance and diversity of insects. The role of organic farming in mitigating the impact of agriculture intensification, urbanization, deforestation, and climate change on global insects' decline and diversity loss is discussed.


**Keywords:** insect biomass, biodiversity, ecosystem, organic farming, insect decline, landscape heterogeneity

## 1. Introduction

Compared with vertebrates, insects had not been given much more attention with respect to loss of diversity and conservation [1]. Recently, entomologists in Krefeld city in Germany published an article reporting a 76% decline in insects' biomass in a study that extended over 27 years [2]. This study "Krefeld study" has sparked a lot of global discussion among insect scientists as well as in the public media. Alarming terminologies were used to describe the event such as ecological Armageddon, insect Armageddon, insect defaunation, insect apocalypse, and insect decline in the Anthropocene. The Krefeld study has become connected with global insect decline as "silent spring" is connected with the negative impact of pesticides. Another study conducted by Lister and Garcia [3] in Mexico in rainforest over 36 years reported a decline of 98 and 78% for epigeal and canopy-dwelling arthropods, respectively. Sánchez-Bayo and Wyckhuys [4] performed a meta-analysis on 73 reports on insect decline all over the world and reported a drastic decline that may lead to the total loss of 40% of the world's insect species. These alarming indicators of global insect decline led many researchers to try to find the causes and the consequent impact of this decline on the ecosystems. The main causes of insect decline appear to be habitat loss, conversion to intensive agriculture, urbanization, invasive species, climate change, and pollution by synthetic pesticides and fertilizers [4, 5]. Of the abovementioned possible causes, agricultural intensification and habitat loss are the main causes of global insect decline [2, 6–8]. Habitat losses are mainly through the removal of forest covers, urban expansion, light pollution, and industrialization, which is responsible for polluting terrestrial and aquatic environments of arthropods [5]. The overall impacts of global insect decline on the proper functioning of the ecosystem could be easily manifested through the decrease in the services that the ecosystem provides in terms of pollination, trophic interaction, and nutrient recycling [9]. Maintenance of insect habitats, cut in synthetic pesticide use [10], and organic farming [11, 12] are probably the most effective means to stop a further decline of insects and promote recovery of biodiversity. This chapter aims to summarize the possible causes of global insect decline, the impact on ecosystem services, and measures to alleviate it with emphasis on the organic farming system.

## 2. Role of insects in the ecosystems

The total number of insect species in the world is estimated to be about 1 million with approximately 4.5–7 million remaining to be identified and named [13]. Insects performed three natural processes, which are essential for the proper functioning of the ecosystem. These are pollination of fruit blossoms, decomposition of organic matter into humus, and natural pest regulation (**Figure 1**) [3, 14, 15].

Insects represent a major source in the food web particularly for birds, reptiles, amphibians, and fish, which represent higher trophic levels. Other invaluable ecosystem services provided by insects include pollination of more than 75% of crops and wild plants [16], waste disposal and nutrient cycling, provision of high-value products such as honey, silk, venom, and shellac. Insects also provide a source of protein for domestic animals and humans (entomophagy) [7, 15]. In the United States alone, the annual ecosystem services provided by wild pollinators were estimated at $57 billion (**Figure 2**) [17]. The relationship between the diversity of pollinators and plants in an area is reciprocal. An unbalanced diversity of pollinators may lead to unbalanced plant diversity due to certain plants being selectively pollinated. Thus, the diversity of wild bees strongly influences the diversity of weeds

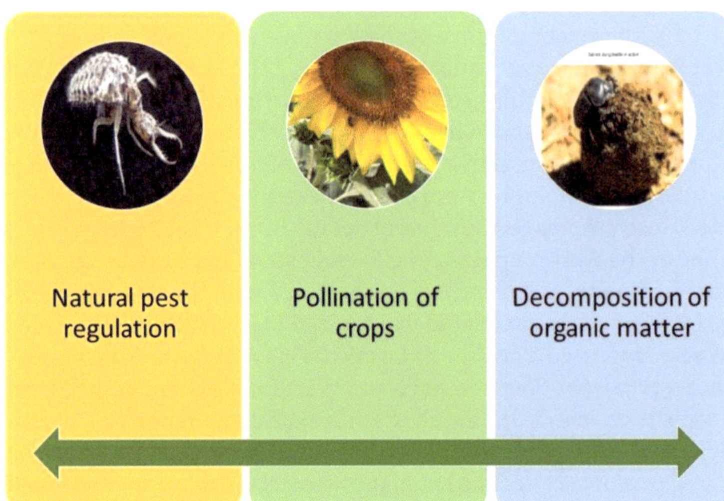


**Figure 1.**
*Main ecosystem services provided by insects to maintain resilience, sustainability, and proper functioning.*

**Figure 2.**
*Bees and butterflies have a significant role in the ecosystem as pollinators.*

and *vice versa* [18]. It is worth mentioning here that Garibaldi et al. [19] reported that the conservation of bees' diversity is essential for ecosystem biodiversity. Other pollinators, which are important but overlooked, include hoverflies (Syrphidae). They perform different ecological functions such as pollination of a wide range of plants, controlling insect pests as biocontrol agents, and being used as bioindicators for monitoring the ecosystem's functioning.

Insects are the *sine qua non* for proper functioning ecosystems that also provide intangible services such as collection for recreational and esthetic values [15, 20]. Understanding the significant role of insects in the well-being of the planet by the public will greatly help in the adoption of mitigating measures that at least decrease the rate of decline of this group of animals. In this respect, increasing the awareness of people about the significant ecological role of insects as pollinators, prey, and nutrient recyclers could be achieved through community (citizen) science and other extension media [15].

## 3. Possible causes of global insect decline and its impact on ecosystems

Destruction of insect habitat, agricultural intensification, urbanization, invasive species, agro-chemical pollution, and climate warming are the main causes of global insect declines and loss of biodiversity [4, 14, 21]. Climate warming is important in the tropics; however, it may have a limited impact on the number of species in temperate regions [4]. Agricultural intensification, urbanization, deforestation, and pesticide pollution account for about 78.7% of the decline causes, while other drivers such as invasive species, climate warming, and other pollutants account collectively for only 21.3% [4]. Destruction of insect habitat is one of the important anthropogenic activities, which is responsible for biodiversity loss (**Figure 3**) [4].

Pesticide use is considered an important cause of global insect decline and biodiversity loss [22]. Consequently, insect decline indirectly influences vertebrate predators [22]. Herbicides, which are extensively used in conventional agriculture largely, eliminate weeds and wild plants, which provide a source of food and shelter for arthropods, both pests and their natural enemies. Changes in insect biomass are more relevant for the ecological functioning of the ecosystem [2].

**Figure 3.**
*A feral colony of the dwarf honeybees,* Apis florea *on a newly cut branch of the button mangrove,* Conocarpus erectus *L. (type of habitat destruction).*

A seasonal decline of 76% of the flying insect biomass was reported to have occurred in Germany during 27-year study of continuous insect monitoring using Malaise trap [2].

Aquatic invertebrates including crustaceans, mayflies, caddisflies, and dragonflies are very much affected by pyrethroid insecticides. Neonicotinoids, on the other hand, affect pollinators including honeybees and bumblebees, particularly when used as a post-bloom spray on perennial trees and field crops.

Industrial pollution is among the important causes of global insect decline, and the fertilizer industry may account for 10% [4]. Light pollution can lead to the luring of moths to bulbs, and make insects fall prey to lizards, toads, birds, and other predators. This negatively affects insects that use their own body-produced light as signals for mating as in the case with fireflies. Mercury vapor, metal halide, and compact fluorescent bulbs induce a more negative impact on moths (sensitive to artificial light at night) than LED and sodium lamps [23]. However, the effect of artificial light on insect populations and declines remains to be elucidated. In their assessment of the drivers causing global insect decline, Sánchez-Bayo and Wyckhuys [4] reported that *Lepidoptera*, *Hymenoptera,* and *Coleoptera* (dung beetles) in terrestrial ecosystems and *Odonata*, *Plecoptera*, *Trichoptera*, and *Ephemeroptera* from aquatic taxa were more affected.

Insect biomass has been used as a proxy for measuring the biodiversity of insects; however, this index has its limitations [24]. Instead, they recommended robust measures of biodiversity trends based on metrics including traits-based phylogeny according to spatial and temporal changes. Additionally, Didham et al.

[25] emphasized the inclusion of data from long-term studies and diversity metrics in the measurement of insect decline. To reach a consensus on the global decline of insects, more log-term studies of biomass, abundance, and biodiversity are needed [26]. In tropic and subtropics, where the majority of insect diversities exist, there are few or no records of long-term data and checklists for most of the species as the case in a temperate region. Thus, many of the species may go extinct in the tropics without being noticed and in most cases before being identified and named. Due to the abovementioned reasons, the impact of global insect decline on the proper functioning of the ecosystem is yet to be quantified [27]. Therefore, long studies and compilation of records and checklists are urgently needed in the tropics and subtropics.

## 4. Differences between conventional and organic farming systems

In the organic farming system, natural biological processes such as the activities of soil microorganisms, nutrient cycling, and biocontrol agents are used in pest management to keep pest populations below the level that cause economic damage. On the other hand, tillage and cultivation practices are used to manage soil fertility and crop nutrients [28–31]. This is contrary to what happens in conventional farms, where synthetic chemicals including fertilizers, insecticides, and herbicides are commonly used in pest management. Pest management in organic farms is carried out by using mainly botanical and microbial pesticides that are either harmless or with a little adverse effect on the agroecosystem. Other options of pest control include crop rotation, mechanical cultivation, mulching, and flaming. Due to the use of benign pesticides and other environmentally friendly pest management measures, organic farms have a high diversity of arthropod species, on average, than conventional farms [30]. Organic agroecosystems differ from conventional ones by greater insect diversity [32], as indicated by the relevant indices, as well as the diversity of taxa and the number of individuals [12]. The largest number of phytophages was recorded in the organic fields of winter wheat, but in organic ecotones and adjacent protective forest shelterbelts, compared with the conventional ones, there were a larger share of predators and parasites. The similarity of organic field ecosystems and conventional forest belt by the Sørensen coefficient indicates the migration of phytophages from conventional fields to adjacent areas [11].

## 5. Impact of organic farming on faunal and floral biodiversity

Biodiversity encompasses different levels including species diversity, genetic diversity, and habitat and ecosystem diversity. It is essential for proper ecosystem functioning and critical processes such as pollination, reduction in soil erosion on arable land, decomposition of dung in pastures, and natural pest reduction in soil and on crops. Biodiversity is also essential for the stability and resilience of ecosystems [24, 33].

Species richness is higher in organic agriculture and pastures than conventional ones because chemical veterinary drugs do not contaminate them. Dung beetles provide an essential ecological function by degradation and recycling of dung, which add to soil fertility and quality in natural or organic pastures (**Figure 4**).

Dung beetles encompass three groups; the rollers (Scarabaeidae), tunneller (Geotrupidae and Scarabaeidae), and dwellers (Aphodiidae) [4]. Organic farming increases the richness and abundance of insects as compared with conventional farming. The number of insect orders, families, and individuals is greater under

**Figure 4.**
*A sacred dung beetle contributing to the recycling of nutrients in pasturelands.*

organic farming. This is supported by the meta-analysis of several published studied on the topic [6, 12, 34]. Moreover, biodiversity indices such as Shannon, Menhinick, Margalef, Berger-Parker, and Piclou confirmed the greater diversity of insects in the organic field of winter wheat. The number of predators and parasitoids was more than double in organic ecotones and forest shelters [11]. Insect species richness and abundance in organic farming were found to be 22 and 36% higher than conventional farming. Likewise, the species richness and abundance for spiders were 15 and 55%, respectively higher compared with conventional farming [12]. Organic farms provide alternative habitats for predator and parasitoid communities through hedges, which represent refugia and source of food (pollen and nectar) for the adults of many insect species (**Figure 5**). Marshall et al.

**Figure 5.**
*Hedges around organic farms provide refugia for predators and parasitoids.*

[35] confirmed this and indicated that many species of arable weeds support a large variety of insect species.

Schmidt et al. [36] reported higher spiders' densities of about 62% in organic farms than conventional farms. They also highlighted the impact of landscape diversification, which is common in organic farming on parasitoid wasps, ladybird beetles, and ground beetles.

## 6. The main practices on organic farms that promote higher insect biodiversity

Conventional, seminatural or organic and landscapes surrounding the farms as well as the farm size greatly influence the conservation of biodiversity [6]. The practices and crop husbandry measures in organic farms that influence biodiversity include no use of herbicides, forbidden of synthetic chemical pesticides, use of pure organic fertilizers, rotation with a leguminous crop, and heterogeneous farm structure [37]. All these practices increase arthropod diversity, particularly soil-dwelling insects. The organic farming system encourages natural processes such as decomposition of organic, usage of livestock dungs, and compost in which several species of insects can thrive. Saprophagous insects such as springtails (*Collembolla*) flourish well in organic farms than conventional ones. The use of cover crop mulching to increase soil fertility, maintain temperature, and conserve moisture enhances the presence of insects. Soil disturbance is to the minimum in organic farming; thus, soil microorganisms and arthropods can thrive well. In an organic farming system, the use of predators and parasitoids together with botanical or natural microbial insecticides has no deleterious effect on the arthropod communities. Honeybees, wild bees, and bumblebees were reported to have exploited the diversity of different flora in organic fields. Diversity of weeds, trees, and shrubs as well as hedges in organic farms encourages visitation of bees and other generalist pollinators [38].

## 7. Conclusions

The analysis of data on insect biodiversity revealed that organic farming strongly encourages the abundance and biodiversity of insects. Monitoring global insect decline based on biomass as the sole metric is not enough. Most of the studies on biodiversity were carried out in the temperate region, where log-term data exist. A few studies are available in tropical and subtropical regions where the majority of insects exist; therefore, no available data upon which trends of insect decline can be traced. Using robust methodology for monitoring decline in abundance and biodiversity as well as long-term studies, data are needed to reach a consensus on the main drivers of this decline. Organic farming and landscape heterogeneity can be adopted as farming systems that alleviate the loss of insects' biodiversity and decrease the rate of global insect decline.

## Author details

Hamadttu Abdel Farag El-Shafie[1,2]

1 Date Palm Research Center of Excellence, King Faisal University, Al-Ahsa, Saudi Arabia

2 Faculty of Agriculture, Department of Crop Protection, University of Khartoum, Shambat, Sudan

*Address all correspondence to: elshafie62@yahoo.com

## References

[1] Leandro C, Jay-Robert P. Perceptions and representations of animal diversity: Where did the insects go? Biological Conservation. 2019;**237**:400-408

[2] Hallmann CA, Sorg M, Jongejans E, Siepel H, Hofland N, Schwan H, et al. More than 75 percent decline over 27 years in total flying insect biomass in protected areas. PLoS One. 2017;**12**: e0185809

[3] Lister BC, Garcia A. Climate-driven declines in arthropod abundance restructure a rainforest food web. Proceedings of the National Academy of Sciences of the United States of America (PNAS). 2018;**115**(44):E10397-E10406 www.pnas.org/cgi/doi/10.1073/pnas.1722477115

[4] Sánchez-Bayo F, Wyckhuys KAG. Worldwide decline of entomofauna: A review of its drivers. Biological Conservation. 2019;**232**:8-27

[5] Ceballos G, Ehrlich PR, Dirzo R. Biological annihilation via the ongoing sixth mass extinction signaled by vertebrate population losses and declines. Proceedings of the National Academy of Sciences of the United States of America (PNAS). 2017;**114**: E6089-E6096

[6] Montañez M, Amarillo-Suárez A. Impact of organic crops on the biodiversity of insects: A review of recent research. Revista Colombiana de Entomologia. 2014;**40**(2):131-142

[7] Gallé R, Happe A-K, Baillod AB, Tscharntke T, Batáry P. Landscape configuration, organic management, and within-field position drive functional diversity of spiders and carabids. Journal of Applied Ecology. 2019;**56**:63-72

[8] Lehmann P, Ammunět T, Barton M, Battisti A, Eigenbrode SD, Jepsen JU, et al. Complex response of global insect pests to climate warming. Frontiers in Ecology and the Environment. 2020;**18**(3):141-150

[9] Hallmann CA, Foppen RP, van Turnhout CA, de Kroon H, Jongejans HE. Declines in insectivorous birds are associated with high neonicotinoid concentrations. Nature. 2014;**511**:341-343

[10] van Strien AJ, Meyling AWG, Herder JE, Hollander H, Kalkman VJ, Poot MJM, et al. Modest recovery of biodiversity in a western European country: The Living Planet Index for the Netherlands. Biological Conservation. 2016;**200**:44-50

[11] Grabovska T, Lavrov V, Rozputnii O, Grabovskyi M, Mazur T, Plishchuk Z, et al. Effect of organic farming on insect diversity. Ukrainian. Journal of Ecology. 2020;**10**(4):96-101

[12] Stein-Bachinger K, Gottwald F, Haub A, Schmidt E. To what extent does organic farming promote species richness and abundance in temperate climates? A review. Organic Agriculture. 2020;**11**(4):1-12. DOI: 10.1007/s13165-020-00279-2

[13] Stork NE. How many species of insects and other terrestrial arthropods are there on Earth? Annual Review of Entomology. 2018;**63**:31-45

[14] Fox R. The decline of moths in Great Britain: a review of possible causes. Insect Conservation and Diversity. 2013;**6**(1):5-19. DOI: 10.1111/j.1752-4598.2012.00186.x

[15] Kawahara AY, Reeves LE, Barber JR, Black SH. Eight simple actions that individuals can take to save insects from global declines. Proceedings of the National Academy of Sciences of the United States of America (PNAS).

2021;**118**(e2002547117):1-6. DOI: 10.1073/pnas.2002547117

[16] Ollerton J, Winfree R, Tarrant S. How many flowering plants are pollinated by animals? Oikos. 2011;**120**(3):321-326. DOI: 10.1111/j.1600-0706.2010.18644.x

[17] Losey JE, Vaughan M. The economic value of ecological services provided by insects. Bioscience. 2006;**56**(4):311-323

[18] Klaus H, Kleinebecker T, Prati D, Gossner MM, Alt F, Steffen B, et al. Does organic grassland farming benefit plant and arthropod diversity at the expense of yield and soil fertility? Agriculture, Ecosystem and Environment. 2013;**177**:1-9

[19] Garibaldi LA, Aizen MA, Klein AM, Cunningham SA, Harder LD. Global growth and stability of agricultural yield decrease with pollinator dependence. Proceedings of the National Academy of Sciences of the United States of America (PNAS). 2011;**108**:5909-5914

[20] Janzen DH, Hallwachs W. To us insectometers, it is clear that insect decline in our Costa Rican tropics is real, so let us be kind to the survivors. Proceedings of the National Academy of Sciences of the United States of America (PNAS). 2021;**118**(2e2002546117):1-8. DOI: 10.1073/pnas.2002546117

[21] Benton TG, Bryant DM, Cole L, Crick HQ. Linking agricultural practice to insect and bird populations: A historical study over three decades. Journal of Applied Ecology. 2002;**39**(4): 673-687. DOI: 10.1046/j.1365-2664.2002.00745.x

[22] Schulz R, Bub S, Petschick LL, Stehle S, Wolfram J. Applied pesticide toxicity shifts toward plants and invertebrates, even in GM crops. Science. 2021;**372**:81-84

[23] Boyes DH, Evans DM, Fox R, Parsons MS, Pocock MJO. Is light pollution driving moth population declines? A review of causal mechanisms across the life cycle. Insect Conservation and Diversity. 2021;**14**(2):1-21. DOI: 10.1111/icad.12447

[24] Vereecken NJ, Weekers T, Leclercq N, de Greef S, Hainaut H, Molenberg J-M, et al. Insect biomass is not a consistent proxy for biodiversity metrics in wild bees. Biological indicators. 2021;**121**:107132. DOI: 10.1016/j.ecolind.2020.107132

[25] Didham RK, Basset Y, Collins CM, Leather SR, Littlewood NA, Menz MHM, et al. Interpreting insect declines: Seven challenges and a way forward. Insect Conservation and Diversity. 2020;**13**(2):103-114

[26] Habel JC, Ulrich W, Biburger N, Seibold S, Schmitt T. Agricultural intensification drives butterfly decline. Insect Conservation and Diversity. 2019;**12**(4):289-295

[27] Wagner DL, Grames EM, Forister ML, Berenbaum MR, Stopak D. Insect decline in the Anthropocene: Death by a thousand cuts. Proceedings of the National Academy of Sciences of the United States of America. PNAS. 2021;**118**(2):e2023989118. DOI: 10.1073/pnas.2023989118

[28] IFOAM (International Movement of Organic Agriculture Movements). Definition of Organic Agriculture 2018. https://www.ifoam.bio/en/organic-landmarks/definition-organic-agriculture. [Accessed: 20 October 2018].

[29] Hole DG, Perkins AJ, Wilson JD, Alexander IH, Grice PV, Evan AD. Does organic farming benefit biodiversity? Biological Conservation. 2005;**122**: 113-130

[30] Gomiero T, Pimentel D, Paoletti MG. Environmental impact of

different agricultural management practices: Conventional vs. organic agriculture. Critical Reviews in Plant Sciences. 2011;**30**:95-124

[31] Henning J, Baker L, Thomassin PJ. Economics issues in organic agriculture. Canadian Journal of Agricultural Economics. 1991;**39**:877-889

[32] Letourneau DK, Goldstein B. Pest damage and arthropod community structure in organic vs. conventional tomato production in California. Journal of Applied Ecology. 2001;**38**:557-570

[33] Di Falco S. On the value of agricultural biodiversity. Annual Review of Resources Economics. 2012;**4**:207-223

[34] Rahmann G. Biodiversity and organic farming: What do we know? Landbauforschung-vTI Agriculture and Forestry Research. 2011;**3**(61):189-208

[35] Marshall EJP, Brown VK, Boatman ND, Lutman PJW, Squire GR, Ward LK. The role of weeds in supporting biological diversity within crop fields. Weed Research. 2003;**43**:77-89

[36] Schmidt MH, Roschewitz I, Thies C, Tscharntke T. Differential effects of landscape and management on diversity and density of ground-dwelling farmland spiders. Journal of Applied Ecology. 2005;**42**:281-287

[37] Bengtsson J, Ahnström J, Weibull AC. The effect of organic agriculture on biodiversity and abundance: A meta-analysis. Journal of Applied Ecology. 2005;**42**:261-269

[38] Lentini PE, Martin TG, Gibbons P, Fischer J, Cunningham SA. Supporting wild pollinators in a temperate agricultural landscape: Maintaining mosaics of natural features and production. Biological Conservation. 2012;**149**:84-92